ON PRESERVING
TROPICAL FLORIDA

John C. Gifford

ON PRESERVING
TROPICAL FLORIDA

Compiled and with a Biographical Sketch by
ELIZABETH OGREN ROTHRA

Drawings by Jane Gifford Howell

UNIVERSITY OF MIAMI PRESS
Coral Gables

Drawings by Jane Gifford Howell
Illustrations: Historical Association of Southern Florida, pp.
50-51; Fairchild Tropical Garden: 29, 30, 33, 52; all others
from Gifford family archives.

For Dann and Katie

Contents

Illustrations

Drawings throughout are by Jane Gifford Howell

Preface: The Editor's Dilemma

Picking and choosing what to include in a collection of a man's writings is never taken lightly. Reading brief paragraphs by this colorful writer is entertaining, but often it does not convey the scholarship and thought of the complete text. When John Gifford had an idea, he chewed and gnawed at it until he had wrung every last bit of good from it. Take the statement, "Lawns are a northern tradition." Before he reached his subject of tropical yards, he wrote of grass as pasture, essential to survival in northern climates. He described the way it grows, its roots and rhizomes, and its use to man in protecting the surface of the earth against leaching and erosion. He pictured the voracious canebrakes of the South and the useful tree-grass or bamboo of the Orient.

While efforts have been made to place Dr. Gifford's material in the context of its times in Part I, little or no editing or modernization of style in punctuation has been made in Part II, in order to preserve the flavor of the original. This book is intended to introduce the reader to the writings of John Gifford, and it should be remembered that they spanned a period of over 40 years. As such, they reflect the varying styles of the times.

Readers who enjoy this collection of writings will want to dip into Dr. Gifford's complete works. Though out of print, his books can be found in many public libraries and on the shelves or lists of some rare book dealers.

Acknowledgments

This book had its beginning in the vivid reminiscences of Mrs. John C. Gifford of her late husband, and I am most appreciative for all her gracious assistance. Dr. Gifford's daughters, Mrs. Harold Melahn and Mrs. Cotton J. Howell, both spent long hours elbow-deep in the massive wooden chest that held Dr. Gifford's manuscripts, photographs, letters, scrapbooks, notebooks, and other private records. Thanks are extended to them and to Mrs. Gifford for their generosity, both in time and in granting permission to me to use both these materials and Dr. Gifford's published writings.

For sharing their memories of Dr. Gifford with me I am indebted to the late Lewis Mulford Adams; his daughter, Mrs. Harold Smith; Dr. Gifford's daughter by marriage, Mrs. Thomas F. Smith; Charles M. Brookfield, Donald Early, Alfred Browning Parker, Angus B. Harrison, and W. T. Price; also many colleagues and former students at the University of Miami including Dr. Taylor Alexander, Dr. Ernest Miller, Ernest McCracken, Dr. J. Riis Owre, Mrs. Melanie Rosborough, Dr. Charlton W. Tebeau, and Dr. Leonard Muller and Mrs. Muller, née Barbara Fairchild.

I should also like to thank the Historical Association of Southern Florida and their museum director, David Alexander, for access to the complete file of the *American Eagle,* and the loan of photographs and other valuable records; George Rosner of the Richter Library at the University of Miami for his contribution of anecdotes and procuring of rare material for Dr. Gifford's biographical sketch; Julia Morton of the University of Miami for permission to use materials from botanical files; to the librarians of the Coconut Grove Library and the Miami Public Library; and to the Fairchild Tropical Garden and its Director, Dr. John Popenoe, for access to its technical library, the letters of Dr. David Fairchild, and the loan of photographs.

Others helping to compile Dr. Gifford's bibliography included the Hume Library at the University of Florida, the Florida Department of Agriculture, and Mrs. Vivian Wiser, historian of the United States Department of Agriculture. In verifying certain points of early history, the Society of American Foresters, the Forest History Society, the American Forestry Association, George R. Moorhead, forester of the State of New Jersey, the Information Office of Cornell University, and the United States Plant Introduction Station in Miami all had a part.

Most of all, perhaps, I am indebted to the author of the book, John Clayton Gifford, whose warm-hearted personality left such a strong imprint that he never seemed farther than a telephone call away.

Part I

JOHN CLAYTON GIFFORD
1879-1949

by Elizabeth Ogren Rothra

There are many things that render the Florida of old precious to those who have long lived here. The Seminole with his little family in his cypress dugout on creeks, in the glades, the clouds of birds in swamps of cypress or mangrove, quiet lagoons fringed with palmettoes and cocoplums, and last but not least, green keys in particolored waters with the Conch at work in his picturesque boat by his sponge or turtle crawl. These and many other pictures the old-timer will carry in his mind's eye as long as he lives.

Rehabilitation of the Floridan Keys, p. 64

The Overview

John Clayton Gifford has been called South Florida's great interpreter.[1] Certainly he understood the topography of this region better than most men of his time. "Florida, like a long finger, is stuck in the face of the West Indies,"[2] he said, emphasizing the tropical nature of the area. Around him new settlers from the North raised steep roofs on their houses, as though to shed snow, and blasted out cellars in the rocky soil that grew hotter rather than colder the deeper they dug. "Adapt yourself to your new surroundings," he advised them. "Don't expect to change the surroundings to suit you."[3]

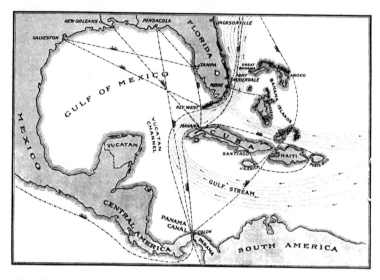

Map from one of Dr. Gifford's articles in Everglade Magazine, *1911. Original caption read: "Map showing Florida's advantageous geographical location—the nearest warm spot to the big eastern and northern markets and the nearest part of the United States to the Panama Canal. Florida has more miles of coastline than any state in the Union—almost equal in length to the Pacific Coast of the United States, exclusive of Alaska. It is hugged on three shores by the Gulf Stream, which moderates the climate both winter and summer."*

As a public speaker, a forester-conservationist, author, teacher, and civic leader he influenced the development of southern Florida from the turn of the century when Miami was almost unknown, through its remarkable years of growth, its setback during the depression years, and the postwar period. His genial "Thank you and goodbye" closed more than 500 broadcasts over Miami's radio station WIOD in the thirties, and his articles on horticulture in the tropics were read in publications throughout the nation.

Florida was his favorite topic, and most particularly that crescent-shaped coastline region curving from Fort Lauderdale on the east coast through Miami and Homestead, across the state, and up the west coast just beyond Everglades City. This "land of no seasons," though temperate according to lines on the map, he claimed as tropical. "Southern Florida is not semi- or sub-tropical, but entirely tropical," he asserted in his speeches. He used the yardstick of the botanist, the zoologist, and the ornithologist who saw the region as the dividing line between temperate zone species and those of the West Indies, Mexico, Central America, and other countries of the Caribbean.[4]

Not everyone agrees with him, especially since killing frosts have been felt in the region. Dr. Gifford knew it was not frost-free after the freezes of 1894, 1895, and 1917, and he said so in his writings. It didn't stop him from claiming his tropical Florida as the best climate in the United States, if not the world, and other weather experts backed him up.[5]

With rare perception he pictured tropical Florida as a community on the shores of the Caribbean, related to this region by climate, flora and fauna, and even customs more closely than to the North American continent. Describing Florida's geographical location as its greatest natural resource, he pointed to its coastline, longer than that of any other state and almost equal to the length of the whole Pacific coast of the United States, exclusive of Alaska.

One of Florida's early environmentalists, along with men such as John Kunkel Small,[6] and later, Thomas Barbour,[7] Dr. Gifford warned of the rapid depletion of Florida's natural resources. "The time to preserve a thing is while it is still plentiful,"[8] he cautioned, telling of entire beds of oysters dug out with a dredge and orchids hauled from the Everglades by the truckload. Long before the word "pollution" was on the lips of every schoolchild, he recognized man's potential for fouling his environment. Observing fish kills in Florida's canals, and streams turning coffee-colored from industrial waste, he wrote in *Living by the Land,* "We are facing a shortage of water fit for human consumption. Clearly the conservation of water is one of our prime problems."[9] As early as 1911 he advised in *Everglade Magazine,* "No sewage should be allowed to enter our canals, streams and bays. This restriction should be placed in all deeds." He didn't mince words in driving home his point, either. He would tell of the man who hung himself on a fertilizer sign. "Not until the odor of the decomposing corpse became overpowering did anyone notice it. It is only when conditions begin to smell bad that we are moved to act."

He addressed groups of all kinds, the Kiwanis clubs, and Rotarians,

farmers' Granges and women's beautification committees, horticultur-
alists and university students. To all he urged the appreciation and
preservation of the natural environment. What's more, the salty humor, booming voice, and warm personality of the man carried his
message in such a way that it was not easily forgotten. "Over in
British Guiana," he would claim, "the natives have to carry specially
constructed umbrellas lest they be knocked unconscious by the ex-
ploding seed pods of the cannonball tree [*Couroupita guianensis*]."
The tree, his listeners found, was genuine. About those umbrellas,
well, that remained to be seen.

His tropical botany students at the University of Miami in learning
about poisonous plants and animals of the region would hear him say,
"If you are bitten by a coral snake, find a comfortable spot and lie
down. You might just as well enjoy your last twelve minutes of life.
However, this 'dangerous' coral snake is only as big around as your
finger and not more than nine inches long. It has one long fang which
holds the poison, and to sink that fang, it must open its mouth way
up. Furthermore, it is nocturnal and seldom found anywhere but in a
rubbish pile. Anyone who will stand barefoot on a rubbish pile at
midnight long enough for that snake to deliver his poison, well, damn
it, he deserves to die!" Little wonder that Dr. Gifford packed his
classes not only with students but also with visitors who just dropped
in to see the show.

He liked to tell a good story, but between the laughs he mixed the
great lessons of man's interdependence with his environment. "He
who scorns nature's bounties will in time suffer her furies," he said. [10]
"He was years ahead of his time in predicting the present ecological
crisis," says his colleague, Taylor Alexander of the University of
Miami Department of Biology.

Considered America's first graduate forester, he received his Doctorate in Forestry from the University of Munich in 1899. He was
prominent in the early forestry movement in the state of New Jersey
and an assistant professor at Cornell University's first forestry school,
1899-1903, participating in Dr. Bernhard Fernow's controversial demonstration forest at Axton. After the age of thirty, however, he did
not practice forestry in the sense of managing tracts of woodland. It is
for his role as a communicator of forest ecology, forest economy, and
conservation as it related to tropical Florida that he was primarily

known. He devoted his life to studying and experimenting with trees, demonstrating their role as soil builders, windbreaks, bearers of food crops, and valuable wood products as well as being an essential ingredient of human life. "There are substitutes for the products of the forest, but no substitutes for the forest," he said.

An international authority on trees, Dr. Gifford knew the scientific and common names of countless trees and plants with special reference to southern Florida. As a witness in ecological suits under the Arkansas Swamp Act, he once gave the scientific names for so many species for a court in Jacksonville that the stenographer begged for mercy! His reputation for wood identification was known by botanists throughout the world. He could name a tree from its wood or bark almost as easily as he could from its flower or leaf. Once when beachcombing he found an unusual piece of wood and was stumped on its name. He sent it to an expert in New York for identification. The New Yorker puzzled over it and passed it on to the best authority on wood he knew, the United States Department of Agriculture, who in turn sent it on to their authority on wood identification, John C. Gifford!

Like David Fairchild, H. L. Nehrling, and Charles Torrey Simpson,[11] he was a true plantsman, experimenting for his own pleasure with new varieties, ever seeking species tolerant to tropical Florida. He regarded nature as the great experimenter, however, and had a hearty guffaw for his northern friend who refrigerated young peach trees to simulate frost and encourage their growth in the tropics.

John Gifford made such a strong impression on those he met that years after his death, people tend to slip into the present tense when talking about him. "He was a big man, physically and in every way," recalls Charlton W. Tebeau of the University of Miami Department of History. "We all talked to him whenever we had a chance. He had a far-reaching mind."

His wife, Martha, describes his hands. "They were so large he could hold a dozen eggs in them." They were useful hands, able to wield an ax or a dibble or to guide the tiller of a boat. He liked working with his hands, and prided himself on his technique in nailing up shingles. Though he used but one nail apiece, he seldom lost a shingle in a hurricane.

Except for a sandy fringe of hair, he had been bald since youth and looked older than his years. Gold-rimmed glasses, vandyke beard, waistcoats, and four-button jackets completed the dignified look of the young forester-professor. As he grew older he cared little for clothes and often ordered them from catalogs. He had a pair of khaki pants he liked to wear on a construction site with his carpenter, Bill Mabry.[12] Round as a stove pipe those pants were, and almost as fireproof. He was a pipe smoker and the ashes burnt holes in his good suits. His wife didn't approve of those khaki pants, especially when she wanted him to dress for a dinner engagement. According to his old friend, Lewis Adams,[13] one time he agreed to wear a dinner jacket but refused to change his khaki pants. "My legs will be under the table anyway, and no one will notice," he grumbled. So the story went.

Viewing him through the mirror of his writings, he was a man who preferred the coco palm to the pine tree. "Nothing sorrowful about a coco palm," he wrote. "In a breeze they never emit a whining tune as do the pines, but a lusty clattering and banging."[14] Nothing sorrowful about Dr. Gifford either. He made a million dollars and lost it, but took the reversal philosophically. "He always told about it as though it were an amusing story," said Dr. Tebeau.

As a conservationist, he mourned the coming extinction of the ivory-billed woodpecker and feared that the pileated woodpecker would soon follow. "Pileated woodpeckers always bring activity and noise. I find the greatest pleasure in seeing a group of pileated woodpeckers,"[15] he wrote. Not unlike a woodpecker was Dr. Gifford, hammering and banging away in his "classroom voice," using every art of the skilled raconteur to imprint his message that those who abuse nature's laws will suffer for it.

When John Gifford tried to understand something, he studied its beginnings. "Look in the eyes of an Indian, and you look back through the centuries when his ancestors ruled this land [Florida], when he traveled in canoes and traded in beads and pots."[16] By studying the Indian's past, he sought to understand southern Florida's role. So, to understand John Gifford, one must go back to beginnings.

The Boy and Emerging Man

John Clayton Gifford was born in Mays Landing, New Jersey, February 8, 1870, to Daniel and Emily (Frazier) Gifford, the youngest of four children and their only son.

His father, known to his friends as "Fox" Gifford, was a sea captain. Together with two partners, he operated a shipping firm called "The Three Friends." John Clayton was the namesake of one of the partners who became so fond of the boy that he gave him a share of his profits. So John always had a little money. "It was a source of great pleasure to him," says his wife, Martha Wilson Gifford, explaining that he sometimes used it to buy some trinket for his sisters or to satisfy his boyish needs. In later years, it enabled him to make the career of his choice and acquire a good education.

John's earliest memories were of ships and forests. Until he was twelve he spent more time on board ship than most boys do in a lifetime. When his father retired from the sea John helped on the small farm where they lived in Mays Landing, not far from the busy harbor of the Great Egg River. On weekends and holidays he helped in the cranberry bogs to earn a few extra dollars. The woods surrounding his home were his tramping grounds, and he spent many hours there learning woodlore, Indian crafts, and developing an interest in botany. He was encouraged by the minister of the local Presbyterian Church, a Dr. Peters, who was an amateur botanist and often took John with him on collecting trips to help with the spade and footwork. Sometimes they were accompanied by another boy, Francis E. Lloyd, later a professor of botany at McGill University. Early in the morning they would get up and canoe down the Great Egg River, often beating off clouds of mosquitoes. Dr. Gifford later recalled those New Jersey mosquitoes when he met their match in the swamplands of Florida. Dr. Peters taught the boy the botanical and common names of the plants and helped him to mount specimens of bark, leaf, wood, and flowers.

Nearby were the New Jersey "pine barrens," a desolate stretch of sandy woodland once covered with giant stands of virgin pine, cedar, and oak. All had been stripped by loggers for pitch, charcoal, and timbers for ships and homes long before John Gifford had been born. Now little but scrub pine survived. Fire frequently swept the barrens, consuming everything in its path, including farm buildings and homes, as well as depleting the soil. The sight of this wasted earth cut deeply into Gifford's youthful conscience. Later he was to be instrumental in forming policies for protection of New Jersey state forests and control of the fires.

His boyhood homelife was tranquil, says Lewis Adams. "The Giffords were a peaceable family. His mother was most kind, good to the neighborhood children, and never said a harsh word about anyone."

John attended public school in the village and his knowledge of botany was known and respected by his friends and teachers. While a high school student he discovered a new variety of oak tree which was named for him, *Quercus giffordii*. He often said, however, that he was not sure it was a distinct species, but, as a young man, he thought it was and his discovery was credited by scientific authorities of the time.

In 1886 he applied for admission to Swarthmore in southeast Pennsylvania, a small college which had been founded by the Friends just three years earlier. His application was not immediately accepted, however. The accreditation of his village school and his irregular attendance due to sea voyages caused some doubts about his qualifications. An interview was arranged for him and the admissions committee. "So you think you know something about botany," began the questioner, setting before him a large group of plant specimens including a rare fern found only in New Jersey. Asked to give the common and Latin name of each one, to his elders' surprise the sixteen year old youth knew each one. Finally he confessed that the test was really unfair. He had mounted the specimens himself. The identification tags were in his own handwriting. The college collection was the same one given by Dr. Peters from among the specimens they had collected in the New Jersey woods.

John Gifford was accepted into the Swarthmore freshman class and registered in the School of Architecture. His uncle was a builder and had taught him what he could about construction. The idea of building homes for people appealed to John and, professionally, offered a more lucrative future than the nebulous career of "botanist." Early in his school life, however, he met a very persuasive personality, Dr. Charles Sumner Dolley, a noted professor of botany, who recognized the boy's natural ability and encouraged him to switch to biology, and finally back to his strong suit, botany. Dr. Dolley continued his interest in John Gifford, who wrote later of many pleasant evenings spent with Dr. Dolley and other Swarthmore students at the professor's home in Philadelphia.

In his junior year he transferred to the University of Michigan in

Ann Arbor, where he took advanced work in botany including special studies in fungi, vegetable physiology, and the plant cell. His studies there also included macroscopic petrography and paleontology in the Geology Department. When he elected to return to Swarthmore for his degree, his professors wrote him high recommendations with some dismay. "I was sorry to hear of your determination to leave the University as I had calculated that your intelligence and enthusiasm were destined to bring credit to the department as well as to yourself," wrote Alexander Winchell, professor of geology and paleontology. V. M. Spalding, his botany professor, wrote of his "marked ability, earnestness and conscientiousness," concluding, "I greatly regret his leaving this university, but am assured that his work will be of high order wherever it may be done."

Gifford's bachelor of science degree was awarded in 1890 by Swarthmore. Along the way he had adopted the reasoned, peaceful faith of the Quakers, and he belonged to the Swarthmore Meeting throughout his life.

Upon graduation his parents urged him to consider the medical profession. Few opportunities were open to botanists other than teaching, and even fewer for foresters. In fact, a man had to go to Europe if he wanted a degree in forestry since no graduate courses were offered in America then. The few professional foresters in the United States, like Dr. Bernhard E. Fernow, were Europeans, educated abroad. A national forest policy was not even established until 1885 when Dr. Fernow was appointed the first professional forester to head the United States Division of Forestry. Until this point, the settlers' problems had been clearing trees to make way for crops, rather than the wise management of forests to preserve them for future generations.[17]

Gifford turned his back on botany again and enrolled at Johns Hopkins University (1890-1891), studying pathological histology and bacteriology with special reference to its botanical aspects. He was, again, highly recommended for his work by his professor, William H. Welch.

Next year, however, he was back at Swarthmore teaching. In the spring of that year, Dr. Dolley became interested in establishing a marine laboratory in the Bahamas and sent John Gifford, then only twenty-two, along with others from Swarthmore and the University of

Pennsylvania to the Bahamas to investigate the possibilities of such a venture. Coincidentally, John Gifford had at this time his first glimpse of the little community called Fort Dallas that would one day be the booming metropolis of Miami. Until the Florida East Coast Railroad reached Miami in 1896, southern Florida could be reached only by boat, wagon trail, or walking the beaches. Few people had even heard of the settlement. George M. Barbour's book, *Florida for Tourists and Invalids,* published in 1883, in all its 310 pages never even refers to Fort Dallas (present day Miami). Though the author boasts of traveling in "southern Florida" he never ventured beyond Tampa except to take an excursion to Key West aboard the steamer *Lizzie Henderson.*

Gifford boarded that same steamer on his return trip from the expedition to the Bahamas for Swarthmore and tells about the experience in a report on the Florida Keys [18] written some years later.

"When my job was finished I shipped on an old steamer called the Lizzie Henderson, bound for Tampa with a load of rock for the construction of the Tampa Bay Hotel. We stopped on the way at Key West for fuel and water. At that time there were many people living on the keys, but the mainland was sparsely settled. When we passed Biscayne Bay the captain said there were two small settlements near the Miami River. He said that someday they had hopes of building a city that would be bigger and better than Key West. Unlike Key West, he explained there was plenty of good fresh water there. Little did I realize that that mysterious shore would be my home."

Gifford did not stay long at Swarthmore. He took up his medical studies again at Tulane University, remaining there for two years. On the last leg of his work there he faced up to a growing distaste for a doctor's life. The story of his break with medicine was one Dr. Gifford loved to tell, and an interviewer talking to his friends hears it over and over in slightly different versions. According to Lewis Adams, while serving his internship young Gifford was assigned to the boil ward of Charity Hospital in New Orleans, and every sailor and bum on the waterfront seemed to have boils and carbunkles. Gifford had become the hospital's expert at this unpleasant duty and they found it convenient to keep him at it. One after another his patients filed in and displayed their ugly sores. Gifford would slash the boil, swab it with sulphur, bandage it up, and it was on to the next man. Little wonder his interest in medicine dimmed. Lewis Adams adds that Dr. Gifford was "tender-hearted" as well and didn't like to see a person in pain. He was resentful too, because he had not been promoted.

One evening when sitting on the waterfront watching the busy harbor and thinking over his situation he was approached by a rough looking character. "You may not remember me, but I was one of your patients," he said in introducing himself. Gifford was a sociable person and so he turned to the man and they started to chat. He confessed that he was fed up with the hospital duty and thinking of quitting. The friend proved to be captain of a ship bound for South America. "Come along with me. We haven't a medical man on board and could use your services," he said. John Gifford needed little persuasion. He wrote his resignation to the hospital that night, along with a letter to his mother explaining his decision, and sailed the next day.

Dr. S. P. DeLamp, the hospital surgeon, in writing young Gifford's recommendations, said, "Mr. Gifford was for ten weeks an assistant in my surgical clinic and proved himself a valuable and efficient assistant. Of his work in other departments I have heard but words of praise. Mr. Gifford is a thorough gentleman, courteous in his manners and kind of heart, a careful and intelligent observer, a diligent and ardent student of nature and man. No doubt a brilliant future awaits him in his chosen profession."

Gifford's voyage took him to the countries of the Caribbean, and at each port of call he learned what he could about the people and the land. On many of the islands of the West Indies he found established botanical gardens such as that on the island of Jamaica, where he stood in speechless wonder at the marvelous trees which had been assembled there by their British caretakers. "The sight turned the tide of my life to the tropics as a permanent abiding place," he confided later in a letter to George E. Merrick,[19] regarding the establishing of forest tree gardens in the Bahamas.

While at sea he received a letter from Dr. Charles DeGarmo, then president of Swarthmore, asking him to return to his alma mater and teach botany. Once again he took up his lifelong interest and headed for home to accept the position. He never returned to his medical studies, though they left a deep impression on him. Characteristically, he expressed it by writing about preventive medicine through good eating habits, a kind of human conservation. He championed food medicines like the elderberry and the papaya, and avoided hospitals and doctors throughout his adult life.

Another destiny awaited him at Swarthmore. He was invited to address garden groups, women's clubs, and civic organizations. The popular young professor spoke entertainingly of New Jersey's Indians, about the problems of the brush fires, and the need for establishing wise conservation policies as well as national parks. A handsome man, just under six feet tall with sandy hair and penetrating blue eyes, he soon filled his calendar with speaking engagements. One evening when addressing a women's group in Plainfield, New Jersey, he was introduced to Edith Wright, the attractive and intelligent daughter of an officer of the Red Star Shipping Line who had traveled extensively in the tropics. Though Edith was fifteen years his senior, the pair found much in common and, after a brief courtship, they were married.

Forester in New Jersey

Dr. Gifford is often referred to as "New Jersey's first forester." Officially, no such post existed until 1907 when Alfred Gaskill was appointed. John Gifford, however, was an important figure in the early forestry movement in New Jersey before Gaskill's appointment. The Trenton *True American* notes in an article encouraging young men to become foresters, "With the notable exception of John Gifford, formerly of Mays Landing, whose work for New Jersey and the Cornell University has given him an international reputation, no young man of this state has won prominence exclusively upon his work as a forester."

In the 1890s Gifford donated from his private property the first forest reservation in the state of New Jersey. He surveyed forests in New Jersey at the direction of the state legislature, working closely with Gifford Pinchot and Henry Graves, two important pioneers in forestry and conservation. Gifford's studies in 1898 of the conditions causing destructive fires in the pine barrens of southern New Jersey and his correspondence with the state geologist at this time were contributing forces to the passing of New Jersey's first forestry legislation in 1905. These bills established the Forest Park Reservation Commission with powers to acquire land as state forests, to manage them, and to appoint fire wardens for protection of the woodlands.[20]

Dr. Gifford's report, "The Silvicultural Prospects of the Coastal Plain of New Jersey" was published in 1900 in the Annual Report of the State Geologist of New Jersey for 1899. Eighty-three pages long with nineteen photographs, it contained not only a description of the New Jersey coastal plain and suggestions for wise use of it, but also a comparison to parts of Europe.

In 1895 Gifford founded *The New Jersey Forester,* a publication designed to unite and inform those interested in the need for conservation of the nation's forests. Lewis Adams, his boyhood friend, then an apprentice printer, set the first issue in type. It carried articles by two noted foresters, Bernhard Fernow and Filibert Roth, the latter of whom became Dean of the Michigan Forest School. Under Gifford's capable management, *The New Jersey Forester* developed into an organ of influence far beyond the bounds of its original scope and local character. In 1898 the publication was purchased by the American Forestry Association as their official publication with Gifford retained as editor. It was the predecessor of *American Forests,* the present publication of the association.

As John Gifford delved into forestry he discovered that there was much he needed to know. Since American universities did not offer graduate work in forestry at this time, he decided to study at the University of Munich in Germany. With Edith he sailed for Europe in the summer of 1896. In his pocket he carried a letter to the "Representatives of the United States in Europe" from John W. Griggs, governor of New Jersey, describing him as "John C. Gifford of the Geological Survey of New Jersey and editor of *The Forester,* visiting Europe to study the various systems of forest management and to collect data by personal observation in the field of forest protection, reforestation of wastelands and timber culture which may be applicable to the conditions of the forest lands of New Jersey and may serve for a more efficient protection against fires, for improved methods of timber culture and the general advancement of the cause of forestry in our country."

The "distinguished and most learned man and gentleman, Johanni Gifford" as the magnificently worded Latin document put it, was awarded his doctorate (D.Oec.) from the University of Munich in 1899 and so pleased his examiners that they gave him the key to the city as well. The document was of heroic proportions, sixteen by twenty inches. It described his "most eminent inaugural" dissertation as "The Forestry Conditions and Sylvicultural Prospects of the Sand and Swamp Lands of Southern New Jersey."

It couldn't have been a more opportune time for him to receive his

SUB AUSPICIIS GLORIOSISSIMIS

AUGUSTISSIMI AC POTENTISSIMI DOMINI DOMINI

OTTONIS

BAVARIAE REGIS

COMITIS PALATINI AD RHENUM BAVARIAE FRANCONIAE ET IN SUEVIA DUCIS CET.

IN INCLYTA UNIVERSITATE LUDOVICO-MAXIMILIANEA

MONACENSI

RECTORE MAGNIFICO

ILLUSTRISSIMO ET EXPERIENTISSIMO VIRO

EUGENIO EQUITE DE LOMMEL

*PHILOSOPHIAE DOCTORE PHYSICAE PROFESSORE PUBLICO ORDINARIO INSTITUTI PHYSICI CONSERVATORE REG. LITER. ACADEMIAE MONAC. SOCIO ORD. ORDINUM MER. CORONAE BAVARICAE AC S. MICHAELIS CL. I EQUITE CET.

EXPERIENTISSIMUS ET SPECTATISSIMUS VIR

HENRICUS MAYR

OECONOMIAE PUBLICAE AC PHILOSOPHIAE DOCTOR SILVARUM COLENDARUM PROFESSOR PUBLICUS ORDINARIUS CET.

FACULTATIS OECONOMICO-POLITICAE P. T. DECANUS ET PROMOTOR LEGITIME CONSTITUTUS

PRAECLARO ET DOCTISSIMO VIRO AC DOMINO

JOHANNI GIFFORD

PRINCETONENSI

EXAMINIBUS RIGOROSIS SUPERATIS

DISSERTATIONE INAUGURALI **EGREGIE** SCRIPTA TYPISQUE MANDATA

„THE FORESTRY CONDITIONS AND SYLVICULTURAL PROSPECTS OF THE SAND AND SWAMP LANDS OF THE SOUTHERN NEW-JERSEY"

DOCTORIS OECONOMIAE PUBLICAE GRADUM

CUM OMNIBUS PRIVILEGIIS ATQUE IMMUNITATIBUS EIDEM ADNEXIS

DIE X MENSIS FEBRUARII MDCCCXCIX

EX UNANIMI ORDINIS OECONOMICO-POLITICI DECRETO CONTULIT

IN HUIUS REI TESTIMONIUM HOC PUBLICUM DIPLOMA SIGILLIS MAIORIBUS REGIAE LITERARUM UNIVERSITATIS ET FACULTATIS OECONOMICO-POLITICAE ADIECTIS FACULTATIS EIUSDEM DECANUS ATQUE RECTOR MAGNIFICUS UNIVERSITATIS IPSI SUBSCRIPSERUNT.

University of Munich diploma, February, 1899

doctorate. In 1898 the College of Forestry at Cornell was created by the state legislature of New York. It was the first university school of forestry in America. Funds were also appropriated for the new college to purchase land in the Adirondacks for a demonstration forest. Dr. Fernow, then chief of the Division of Forestry in the United States

Dr. Bernard E. Fernow, Forestry Director and Dean of the Faculty, and Assistant Professors Filibert Roth and John C. Gifford (2nd row from front, persons 3, 4, and 5), with one of the early forestry classes at the first College of Forestry (1899-1903) at Cornell University.

Department of Agriculture, was appointed Director and Dean of Faculty. Gifford Pinchot succeeded him as Chief of the Bureau of Forestry. Dr. Fernow lost no time in gathering his staff. Even before returning from Europe, Dr. Gifford had received an appointment from the Executive Committee of Cornell as Assistant Professor of Forestry at a salary of $1,500.

He accepted the post with great enthusiasm and began teaching in September 1899 courses such as Forest Protection, Forest History and Politics, Forestry with Special Reference to Sylviculture, and German Forest Literature. He also served as Forestry Librarian and gave a practicum at the soon-to-be controversial demonstration forest at Axton near Tupper Lake in the northern range of the Adirondacks. Land had been purchased for the demonstration forest from the Santa Clara Lumber Company, and the Brooklyn Cooperage Company contracted with the school to buy logs for staves and cordwood for wood alcohol. French Canadian lumbermen hiked across the nearby border to cut the trees, mainly mature birch, maple, and beech, which up to that time had not been extensively logged in the Adirondacks. John Gifford had about twenty eager young college men in his Axton workshop. They worked like beavers with few conveniences. Four or five ordinary buildings left behind by the lumber company served as dormitories and dining hall. Dr. Gifford ingeniously fitted together two sturdy chicken houses to form a classroom. They cleaned the forest of overripe hardwoods, established nurseries, and planted thousands of little conifers. Often they were called out to fight fires accidentally set by campers and hunters. Dr. Gifford spoke warmly of the experience and the relationships between the faculty and the student body. "It was more like a family than a public institution," he said. Of Dr. Fernow he recalled that he was a "proficient forester, a good lecturer in at least three languages, and a musician. After a hard day he would play the piano until late at night."[21] The days at Axton flew by, and Gifford was happy to be applying under such compatible leadership the principles of forestry he had learned in Germany. At thirty, he had an international reputation in forestry. As editor of *The Forester* and a university professor he accepted many invitations to speak, among them a series of lectures at Chautauqua Institution in 1900. The following year he was invited to speak in New Hampshire for the Society for the Protection of New Hampshire, and call for the establishment of a national park and forest reservation in the White Mountains.

His first book, *Practical Forestry* (D. Appleton & Co.), was published in 1902. John T. Rothrock, another pioneer forester, reviewed the book for *Forest Leaves*[22] (1902), calling it "the first completed practical treatise on the art and science of forestry published in the country." He highly recommended it both for students and laymen. *Country Gentleman* of December 3, 1902, in reviewing the book, speaks of Dr. Gifford's "discursive style" and his chapters that "have the charm of essays." The book was widely hailed as a valuable addition to America's forest literature, and for many years it was used as a college textbook.

Coincidentally with his writing and teaching at Cornell, Dr. Gifford served as a special agent for the Bureau of Forestry of the U.S. Department of Agriculture. He traveled widely for the bureau in the West Indies, South America, Central America, the Yucatán, and

John C. Gifford, about 1900

Mexico. When a yellow fever epidemic broke out in Cuba (1902) he was sent there to study ways of using trees to drain the swamps where disease-bearing mosquitoes bred. Embarking for Cuba from Miami, Professor Gifford had his first visit to what he later was to call the "American tropics."

In his report on the Florida Keys he tells of this trip.

"Early in the morning the bright yellow-colored train with a chair car for the well-to-do left for Miami. The train stopped one hundred and eighty-seven times on its journey south. The cars were full of dust and smoke. We finally arrived after midnight. It was moonlight and the upper porch of our hotel looked down on many yards full of tropical trees and fragrant flowers. We asked Colonel Waddell if there was anything of special interest in the neighborhood. He recommended a trip to Coconut Grove."[23]

Lewis Adams remembers this first trip to Coconut Grove too. "Hell, the desk clerk took one look at that vandyke beard and the round eye glasses and pegged the doctor for a city slicker. Miami was a real frontier town in those days. Knife fights, robberies, drunken brawls were common. He just didn't want to see Dr. Gifford get into trouble. That's why he recommended Coconut Grove!" explained Lewis Adams.

Dr. Gifford's report continues to say that they hired an old Negro with a pony and coach and followed the new white coral rock road through the hammock to Coconut Grove and the Peacock Inn. He writes, "We liked it so well that we sent back for our baggage. The bay was filled with picturesque sailboats. Some were bringing in turtle, conchs, and fish; others were loaded with truck and fruit from Key West. Seminole Indians loafed on the lawn. The names of men such as Louis Agassiz[24] were on the register. South Florida was practically an island, separated from the rest of the state by miles of mud and unbridged rivers."

Exploring the Luquillo Forest Reserve

In 1902 Theodore Roosevelt established by presidential proclamation the Luquillo Forest Reserve in Puerto Rico. Dr. Gifford, as an "expert" in the service of the Bureau of Forestry of the U.S. Department of Agriculture, was authorized to survey its 28,000 acres at a salary of $1,800.

The opportunity must have seemed a real "plum" to the young forester, but he was also aware of the hardships of exploring this uncharted wilderness, some of it mountainous and much of it tropical rain forest. He accepted the assignment with pleasure, however, and sailed for Puerto Rico in July 1903. Edith, who accompanied him on the trip, proved an excellent companion and was of great help in preparing his report.

Reading between the lines of his later report[25] to Gifford Pinchot, then chief of the Bureau of Forestry, the hardships of his explorations were apparent. Jungle trails were so overgrown with vines and brush that they could be penetrated only with a machete. Slippery stream beds were often the only trail, and these were quickly cut off by sudden downpours which swelled them to foaming torrents. His skin, fair and sensitive, bled from grass cuts and swelled from insect bites and brushes with poisonous plants. The final two weeks of his six-week expedition he was troubled with tropical dysentery.

For the nation's press, when he was interviewed in San Juan before returning to the United States, privation was never mentioned. With great enthusiasm he described a tropical wonderland of giant trees, some as much as 120 feet tall, wild orchids, feathery tree ferns, waterfalls tumbling over precipices hundreds of feet high, and the valuable natural resources of the region. *The New York World* ran the headline, "Paradise found in Porto Rico. Dr. Gifford Saw Glories of Nature No White Man had Looked on in Four Centuries." The *New York Tribune,* headlining their story, "A New National Park," told how Gifford crossed the El Yunque range at its highest point, finding in the mountains huge aromatic gum trees and beautiful streams where natives panned for gold. "It is not at all improbable that the researches of Dr. Gifford will lead to the establishing of the first national tropical park of the United States within this reserve," concludes the *Tribune* report.

Today, the Luquillo Experimental Forest is a national park, crisscrossed by trails and roads and visited by thousands of tourists yearly, thanks in part to this courageous explorer who went before them.

Back at his Princeton home, Dr. Gifford prepared his report for the Bureau of Forestry on his survey of the Luquillo Forest. In his transmittal letter to Chief Forester Pinchot in September 1903 he wrote cheerfully that he was "none the worse for wear in spite of the dysentery brought on by the wet weather and lack of suitable provender."

Meanwhile, a controversy had sprung up in the state legislature of 1903 over Cornell's demonstration forest. Dr. Fernow had planned the forest as an example of good forest management. Unfortunately, few people in America understood that good forestry involved not only the planting of trees and prevention of fires, but also the removal of mature trees. Dr. Fernow intended to harvest the ripe hardwoods and immediately regenerate the forest by planting pine trees, a system often followed in German forests. Dr. Gifford, newly returned from study in Germany, said of Dr. Fernow, "He was a German, steeped in

German traditions and everything he did would have been done in Europe without question. I had seen how the Germans worked every acre for the purposes it was most fit. The Germans claimed they could not afford parks. In fact, the whole country was a park."28

No sooner had the first ax been laid to a venerable maple tree in the demonstration area than a shout went up from private camp owners. "They're cutting down the old maples! " "Woodman save that tree," was the moralistic cry. Few considered the possibility that an experienced forester like Dr. Bernhard Fernow cut trees judiciously. The camp owners were wealthy and politically influential. They soon made their voices heard in Albany. Finding grounds to attack such a worthy institution as the new College of Forestry was not easy, but someone gained access to the books at Axton and discovered the school was buying oleomargarine for their French Canadian lumberjacks' hotcakes! New York state law required public institutions to buy only New York state creamery butter. Dr. Gifford recalled, "It took the best legal talent in the state to keep us out of jail. They accused us of buying ferets to kill the rabbits that bothered our nurseries. These were 'terets,' the white rings used on harnesses. The final so-called inexcusable crime was when Fernow cut down some old maples and birches that bordered on land owned by wealthy summer campers."27

The governor was urged by the campers to veto the demonstration forest bill. Instead, however, in 1903 Governor B. B. Odell, Jr., vetoed the item of $10,000 in the appropriation bill for 1903-1904 which provided for the support of the College of Forestry in Ithaca. In effect, the governor vetoed the wrong bill. Funds were cut off for Cornell's College of Forestry. The professors offered to work for nothing until the mistake could be corrected, but it had gone too far. Some legislators shrugged it off, explaining, "The forests will never be exhausted in the United States anyway, so no need can be seen to study forestry." Others with more vision expressed dismay. Even Governor Odell suggested that the college be reinstated. No criticism was made of the College of Forestry itself. It was the demonstration forest that had roused strong sentiments. The professors and students soon scattered. Dr. Gifford recalled, "It is to their credit that wherever they went they distinguished themselves. All were proud to have been associated with this first American university college of forestry."28

John C. Gifford, early 1900s, when professor at Cornell

It was not until 1910 that the school of forestry was reestablished by the creation of the Department of Forestry in the College of Agriculture.

The president of Cornell expressed regrets to Dr. Gifford at seeing him leave and hinted that a place might develop for him in another department. John Gifford had other plans, however. He already owned a piece of property in Coconut Grove, Florida. He was discouraged by the abandonment of the college and resented the "short shrift" given to faculty and students. Not only that, but an epidemic

"End of the Trail" completed. This was Dr. Gifford's first house in Coconut Grove. It had a red tile roof. Note the widow's walk perched on the rooftop.

of typhoid fever in Ithaca due to foul drinking water had killed several of his friends and so injured his own health that his doctors warned him he might not survive another northern winter.

Pioneer in Southern Florida

Dr. Gifford needed little encouragement to settle in a warmer climate. He had developed a taste for the tropics on his travels, and the little bayfront community of Coconut Grove appealed to him. The Giffords moved to Coconut Grove, where they had property on Biscayne Bay at the foot of Park Lane. There they built, according to Dr. Gifford's design, an unpretentious but comfortable home which they called "The End of the Trail." Its name referred both to their feeling of being settled down at last and to the Indian trail that passed through their property along the bay to a freshwater spring nearby. Up on Grapeland Boulevard (now Southwest 27th Avenue) Dr. Gifford owned a five-acre grapefruit grove. Because of his poor health

Drilling a hole for dynamite in coral rock, the first step in planting a grapefruit tree for Dr. Gifford's plantation in Coconut Grove in the early 1900s.

following his bout with typhoid fever, he was still on crutches when he came to the "Grove." He thought of himself as semi-retired and intended to cultivate fruit and experiment with tropical plants along with managing his wife's estate.

He wrote, "The little city appealed to me and in less than a year I took root, just as a plant takes root when it strikes congenial soil and atmosphere. I was sick when I came here and soon recovered. Ties to a place that have given one his health are of the very strongest and most lasting character. This, together with a strong liking for and interest in tropical plants, and neighbors of an unusually high and likeable grade settled the question for me aside from business opportunities and countless other advantages."[29]

Coconut Grove counted only a few hundred families in the early 1900s, but its settlement by people of character and education such as Ralph Middleton Munroe[30] had attracted, in turn, citizens with an appreciation for the natural beauty of the little bay with its rising coral bluffs. While some of the early inhabitants were people of wealth, such as James Deering, co-founder of International Harvester, and fame, like William Jennings Bryan, the orator and former secretary of state, many too were merely people of somewhat comfortable circumstances like Dr. and Mrs. Gifford who sought a warm, healthful climate. A few miles through the hammock up the white coral rock road was Miami, still a rough seaport and frontier town. Coconut Grove, on the other hand, was a more gentle community where few people locked their doors and a man could still be chastised from the pulpit for so small an offense as repairing his boat on Sunday. Social values often placed as much weight on what a man was as it did on what he owned, and a Negro bonefish guide or a Seminole Indian[31] commanded as much respect as a businessman.

John Gifford fitted into the community perfectly and his white horse and black buggy were a familiar sight around town. As his experience with the tropics became known, he was consulted on everything—what to plant, what to build, what to buy. Ever the plantsman, he delighted in the opportunity to enrich this frontier land with beneficial and beautiful plant immigrants. He exchanged seeds and cuttings with various private concerns and botanical gardens. Often when he returned from a trip to the West Indies or Central America, his coat pockets would be bulging with small cuttings and seeds. "Many people were doing the same thing," he noted. "There were no restrictions on this exchange so it was common and easy."[32]

In 1904 Dr. Gifford brought cuttings of the *Ficus altissima* or lofty fig tree. He planted them on St. Gaudens Road, and they were one of the natural wonders of Coconut Grove until they were felled by the hurricane of 1926. Dr. Gifford recommended the *Ficus altissima* for roadside plantings in South Florida, and for many years the thick foliage of this broad-spreading fig tree lent cooling shade to South Federal Highway in Miami. When they were hewn down by a Miami "beautification committee" and replaced by more manageable, flowering bushes, many people protested. Actually, Dr. Gifford was not sure himself that the "banyan habit" of the mature lofty fig was suitable for the highway.

In his notebook, a tattered, outdated, calendar pad, he once recorded the names of the trees which he thought he had introduced to southern Florida. They were the *Melaleuca leucadendron* (now known as *M. quinquinerva*), or cajeput; *Ficus altissima,* or lofty fig; *Thespesia*

Dr. Gifford looking at seed pods of the cajeput

grandiflora, a tree hibiscus; *Bauhinia kappleri,* or butterfly flower; *Gliricidia sepium,* or nurse tree or madre de cacao; and the *Erythrina indica,* or coral tree. His original coral tree stood for more than forty years at 3940 Douglas Road in Coconut Grove. Its spectacular blood-red blooms carried the message of South Florida's tropical charms around the world on thousands of postcards. When it fell in a mild windstorm in 1946, the *Miami Herald* carried the story in headlines: "City's Original Coral Tree Topples Over." Fortunately its offspring continue to contribute to the colorful spring display of Florida's flowering trees.

John Gifford is best known for his introduction of the melaleuca or cajeput in 1906. His introduction was probably preceded, however, by the United States Plant Introduction Gardens since their plant inventory notes a cajeput being brought in from France in 1900. This tree was apparently propagated and distributed, but it was Dr. Gifford's introduction that brought attention to the tree. Even David Fairchild, then Chief of the U.S. Office of Foreign Seed and Plant Introduction, referred to Dr. Gifford as the first to introduce this

Dear John Gifford.

Isn't this one of the original cajiputs trees which grew from the seeds you imported from Australia and got Simmons to have grown for you?

Its from your old place in Coconut Grove
i. 26. 34   DF.

Cajeput tree at John Gifford's home, thought then to be the first tree of this species in South Florida. Note from David Fairchild on the reverse of above photo.

shaggy-barked, ornamental tree. However, he found it hard to account for the presence of an eighteen foot specimen already growing near Coconut Grove. "Clear this up! " he noted to himself on the back of a photo of Dr. Gifford and a seedling cajeput.

On the west coast of Florida, the cajeput was introduced by A. H. Andrews, who also once considered himself the first to bring the tree to Florida. Mr. Andrews was editor of the *American Eagle,* a weekly publication of the Koreshan Unity group in Estero, Florida. Horticulture was the special focus of the *American Eagle,* and it carried articles by naturalist H. L. Nehrling, David Fairchild, Dr. Gifford, and others. Mr. Andrews later conceded Dr. Gifford's earlier introduction of this Australian plant immigrant. He wrote, "The facts bear out the statement that the *Melaleuca* or cajeput tree was first introduced by Dr. John C. Gifford of Coconut Grove in the early 1900's. I originally thought that I was the first to introduce this tree in Florida. I never visited the lower east coast until 1915 and it was several years later that I learned that Dr. John C. Gifford had preceded me by several years in introducing the *Melaleuca*." As fellow plantsmen Dr. Gifford and Mr. Andrews became good friends, and the *American Eagle* between 1930 and 1948, when it ceased publishing, printed nearly one hundred and fifty articles by Dr. Gifford.

The United States Office of Foreign Seed and Plant Introduction, under the direction of the noted plant explorer, David Fairchild, had established a six acre plant introduction garden on Brickell Avenue, just north of Coconut Grove, in 1898. Dr. Gifford was a frequent visitor to the garden and writes of the role it played in propagating the melaleuca. "I received about a teaspoonful of cajeput seeds [from Australia]. With the help of Ed Simmonds, an enthusiastic tree specialist and a lovable character, I finally succeeded in sprouting these minute seeds. With Simmonds' help a large number of these trees were distributed. Simmonds had charge of the old Plant Introduction Garden on Brickell Avenue in Miami which introduced many valuable things. These were not only given to early settlers, but old timers always stopped there to chat and get information which he gave freely

Edward Simmonds, botanist in charge of the Old Brickell Avenue Garden of the U.S. Department of Agriculture, inspects the roots of crotalaria plants in 1913.

Messrs. Fairchild and Simmonds wading ashore from a boat; Chapman Field Key from the ocean side, March 19, 1922.

from his long experience in practical horticulture. Located on the main highway on the way to Miami, settlers from the Homestead section stopped there on their way to town. It was a social center as well as an experiment station. In fact, it was there that many people first learned to know many of these tropical things. It was a school for all those interested. Under the able guidance of David Fairchild, it has left its stamp on the South Florida landscape."[33]

Altering the native landscape is no longer popular with some modern-day environmentalists. Both the Australian pine and the mela-leuca, for example, have become so naturalized in the Everglades that they are crowding out native species. The mere mention of "melaleuca" will raise the hair of many park naturalists. However, in the perspective of history, the pressing concerns of pioneer foresters in southern Florida were the rendering inhabitable of a wild land with soil so sandy, rocky, wet, or mucky that it hardly passed for soil at all. They searched the woodlands of the world for trees that were salt-tolerant, able to withstand fire, flood, and drought, easily self-propagated, and able to thrive in soil that varied from dry hardpan in winter to flood conditions in summer. Beyond that, they sought trees that would enrich the land and be useful to settlers, supplying timber, edible fruits or nuts, medicines, and products for the home and industry.

In 1912 Dr. David Fairchild and his family moved to Coconut Grove. Dr. Gifford sometimes puzzled over the coincidence that two men in the same "line of work," as he put it, born the same year, should end up in the same location. The two plantsmen must have looked alike, too, since they were often mistaken for each other. Gifford and Fairchild swapped stories as well as plant specimens and exchanged letters from time to time, addressing each other formally in the manner of the times as "Dear John Gifford," or "Dr. Fairchild."

In a letter written in the twenties, Gifford joked to Fairchild, "They say that the right-shaped hand-made tile is hard to get because there is only one man in Spain with the right-shaped leg. I have been watching bathers at the beach and believe that women's legs are better-shaped for that purpose. The best handmade tile is exactly the shape of a fat woman's leg." One can only speculate on the reaction of the more circumspect Dr. Fairchild to his friend's earthy humor.

Gifford and Fairchild were part of a little band of men in South Florida who worked with true devotion to preserve the region's natural beauty and resources. They had a part in defending the rare stand of native royal palms on Paradise Key from destructive fires. Through the fund-raising efforts of the Florida Federation of Women's Clubs, the land was purchased in 1914 and became Florida's first state park, today incorporated into the Everglades National Park as the Royal Palm section. The first visitor's center is here and the famous Anhinga and Gumbo Limbo nature trails, irreplaceable outdoor "museums" where birds, animals, and plants found nowhere else on the American mainland can be observed in the wild.

In his book, *The World Grows Around My Door,* Dr. Fairchild tells

Entrance to Chapman Field in the twenties

of the fight Dr. Gifford put up in Washington to save the "little garden" on Brickell Avenue from being abandoned. Gifford also used his influence in the nation's capital to help Dr. Fairchild acquire Chapman Field, in Cutler Ridge south of Miami, for a large plant introduction garden. The 200-acre site, now one of the four principal plant introduction stations of the United States Department of Agriculture, was once a noted World War I airfield and, through Dr. Fairchild's efforts, was acquired from the Department of the Army in 1923.

Banker, Land Developer, and Builder

With his penetrating personality and hearty enthusiasm for his adopted state, Dr. Gifford became part of the "oligarchy"[34] that steered the destiny of Miami in its early days. He had a considerable fortune. According to Lewis Adams, at one time Dr. Gifford owned one-sixteenth of the stock of the Miami Bank and Trust, one-sixteenth of the Langford Building in Miami, and one-sixteenth of the Miami Mortgage Company. A man of judgment and responsibility, his counsel was highly respected and sought. He was vice president of the Miami Bank and Trust, and president of the Morris Plan Bank, which he helped to found. Known as "the poor man's bank," the Morris Plan allowed persons to borrow money at standard interest rates with no security other than the endorsement of two friends. The idea appealed to Dr. Gifford since it allowed new settlers with limited funds to buy homesteads and become self-sufficient. "The furtherance of the tropical subsistence homestead has been and I hope always will be uppermost in my mind. For forty years my life has been shaping itself to this very end because it seems to me about the most essential thing that can give life and comfort to the majority of our people," he said.

Dr. Gifford's interest did not stop when the signatures were on the loan, however. His door was always open to anyone who needed help, and he advised many a new homesteader on how to get the most from his land. Charles Brookfield, for many years the Florida state representative of the National Audubon Society, recalls how he bought land on Elliott Key for $8.00 an acre from Dr. Gifford. "He went over the land with me and showed me how to restore the abandoned lime orchard on the property so it would yield a profitable crop. He taught me all he could about developing the land," said Brookfield.

Dr. Gifford held many a man's mortgage, and they didn't always net him a profit. On one occasion, he was told that one of his mortgagees had suffered difficult circumstances and was unable to make his payments. "Oh well," the genial doctor was heard to reply, "He's paid enough. Let's just forget about it."

South Florida was enjoying a steady growth. The Florida East Coast Railroad brought an illustrious list of winter visitors to Flagler's Royal Palm Hotel sprawling along the bay front at the mouth of the Miami River. Attracted to Miami as a seasonal health and social resort, many stayed to build mansions along "millionaire's row" on Biscayne Bay. Roads began to fan out into the countryside, and in 1912 the railroad was completed to Key West, then Florida's largest city. That same year, John Collins' wooden bridge spanned two and one-half miles of water to connect Miami Beach with the mainland, and soon the financial genius of Carl Fisher began to develop Miami Beach, America's answer to Italy's Venice.

Gifford was associated with various land companies at this time, usually as an adviser. Among these were the Sunshine Real Estate Company, the Everglade Land Sales Company, the Elliott Key Lime Company, and the Triangle Corporation. His enthusiasm for southern Florida was limitless. It is said that he was directly involved in bringing ninety people to settle permanently in Florida. He wrote promotional material for the land sales companies and the Miami Chamber of Commerce and contributed to the spirit that was drawing thousands of homesteaders to Florida. "Florida is the Italy; Yucatan, the Egypt; and South America the great, unexplored hinterland of this American Mediterranean,"[36] he wrote in stressing southern Florida's resemblance to the countries to the south. Today he would have drawn a handsome public relations fee for his promotion. Then he profited only indirectly through the growth of an area in which he had a personal financial interest.

In his beloved Model T Ford, Dr. Gifford traveled in the southern part of the state, buying up corner lots in small towns for his private investment. Horse-drawn buckboard took him where the Model T could not go. Though he could have owned a more expensive motor car, he liked the plebian nature of the Model T and always was ready to defend it. Once at a conference in Jacksonville a colleague described with great pride the virtues of his new Pierce Arrow touring car. "Well," Gifford commented dryly, "I don't like such a costly gas wagon. I always drive a Model T. If the darn thing breaks down, I just leave it on the highway and buy a new one."

When it came to buying property, Dr. Gifford had a simple formula which he would impart to anyone who asked his advice. "Buy a piece of property," he would advise. "Divide it in two, and hold on till you can sell half for what the whole cost." His daughter Jane (Mrs. Cotton Howell) says that while he profited from real estate, it wasn't the fortune some were making. "He thought he was making a killing when he cut his land in two and doubled his profit while other speculators were splitting the land four ways or more," she said. A. B. Harrison, president of the Coconut Grove Bank and an associate of Dr. Gifford, recalls talking to Dr. Gifford about a piece of property he had a chance to sell at a profit. When he asked the older man's advice, Dr. Gifford said, "Gus, you can never go broke making a profit."

Dr. Gifford bought and sold many acres of Florida land and had fun doing it. Once he took a party of Texans by boat to see a piece of submerged land in the Everglades. Spotting a brilliantly-colored insect about four inches long, Gifford's prospective land customer exclaimed, "What's that? " "Oh, that? That's a Florida grasshopper," Gifford explained casually. The Texan gazed out at the endless horizon of the Everglades and said, "Any land with this much water and grasshoppers as big as birds is bound to be good!" And the deal was clinched.

Another customer that will never be forgotten was a Navajo rug salesman, W. A. Hunter, from Farmington, New Mexico. Mr. Hunter was working with the Indians there, helping them to help themselves by selling their native crafts. Dr. Gifford, a great exponent of home-based industries, sympathized with Hunter's idea. When Hunter found difficulty in coming up with cash for the land he wanted, Dr. Gifford proposed, "Cover the lot with rugs!" He bartered the land for Indian rugs. Mrs. Gifford recalls that the rugs arrived by the carload, handsome, handwoven, wool floor coverings in traditional Indian colors and patterns. "We gave them away as wedding presents for years," she said. Several of the rugs covered the soft brown brick floors of the Gifford home.

Dr. Gifford subdivided his Coconut Grove acreage and helped lay out the streets. True to his Caribbean loyalties, he named the streets for the islands of the West Indies: Andros, Abaco, and Inagua. Others he named for the aboriginals: Tigertail, Calusa, and Seminole. Unlike most developers of his time who arranged their streets in a neat and monotonous north-south gridwork, Dr. Gifford faced his streets to the bay to catch the prevailing winds off the ocean.

The Builder

Construction was a natural adjunct to land development, and Dr. Gifford built many homes and apartment buildings at this time. "He had a natural instinct for shelter," says Alfred Browning Parker, the prominent Miami architect who was both friend and son-in-law to Dr. Gifford. "It was part of his training as an ecologist, partly just in his being." When Dr. Gifford's old home, which he designed and built in the early 1900s, was threatened with demolition, Parker bought it and converted it for his offices.

Though Dr. Gifford had little formal training in architecture, he had learned the practical skills of building from his uncle in New Jersey and is credited with building a house in his youth that is still standing. It was in the School of Architecture that he first registered at Swarthmore, and though he finally gravitated to natural sciences, he never lost his love for building. Wherever he traveled, he studied the methods of shelter. His scrapbooks contain many pictures of unusual dwellings: a fieldstone cottage in England; a log cabin in Virginia, the conical beehive house of Italian peasants shaped to shed the wind.

During the boom days, Miami boasted little or no regional architecture. Downtown Miami's most elegant buildings sported French turrets or neoclassical influences. Coral Gables developers chose a Spanish theme while others built Italian palazzos. Colonial mansions and pseudo-Greek temples reared stiff columns up against the soft, tropical landscape. Few architects troubled to understand tropical Florida's climate well enough to design houses to suit it. Many mistakenly equated it with the climate of Italy and Spain. Dr. Gifford was the exception. He borrowed ideas from the Cuban *bohio*, the South American *hacienda*, and the Seminole *chickee*. Ever conscious of his environment, he had these suggestions for building in southern Florida. "The natural conditions to be considered are long, dry periods, continuous sunshine for months, very heavy rains and strong winds at times. This calls for tight, cool, solid, low structures."[37]

In 1911 Dr. Gifford wrote this advice for building in the tropics, "Build low of rock and timber and when surrounded by vines and shrubbery, the house appears to grow out of the land."[38] Just three years earlier (1908) an almost obscure Chicago architect, Frank Lloyd Wright, had published his "Principles." Wright's theory of binding the house to the land revolutionized American architecture and was similar to the ideas of John Gifford. Gifford read widely and may have been acquainted with Wright's principles. But more likely, Gifford's ideas were his own, born of experience, observation, and his natural instincts.

Gifford's own home at 2937 S.W. 27th Avenue where he lived during his later years was the perfect expression of his ideas. Wild leather ferns, roselle, shrimp plant, bleeding heart, and purple oyster plant grew thick around the foundation. Coral vine, petrea, and flame vine hung from the lofty trees and climbed over the second story, almost obscuring the outlines of the house. In the entrance passage rough Miami limestone underfoot and dark pine rafters overhead gave the visitor the primeval feeling of entering a cave. The walls were formed of limestone blasted from his own property. If the blast left a hole six-foot deep so much the better. Conservationist Gifford just filled it in with junk and compost, threw in a banana root and soon was picking his own fruit conveniently growing at hand level. The

*Entranceway to Dr. Gifford's home on S.W. 27th Avenue
in Coconut Grove*

house was framed in Dade County pine, which he described as "the
meanest wood on earth to work with." No one would dispute the
claim. Used when green, the wood was so gummy that tools had to be
flooded with kerosene to keep them from sticking. Architect Parker
tells Gifford's surefire recipe for nailing up the lumber. "First drill a
hole. Then run the nail over a bar of soap. Fill the hole with kerosene
and after that, if you are lucky, the nail will sink home." Make no
mistake about it. He had great respect for the wood. He saw buildings
built of Dade County pine endure the ravages of wind, water, sun, and

Dr. Gifford's old home on S.W. 27th Avenue, Coconut Grove, where he brought up his family and lived for nearly thirty years. He designed and built the house himself using native products—Dade County pine and coral rock blasted from the ground.

termites. "Once in," he wrote in 1935, "the nails hold for keeps and many of the old houses built with this wood are sound and still standing."[39] He referred to houses built in the early 1800s in the Keys and abandoned by settlers beseiged by Indians.

Dr. Gifford built his house in the pattern of a cross to give maximum exposure to the bay breezes and the warmth of the low winter sun. Heavy roof overhangs protected against summer's heat and sudden rains. Inside, partial walls divided rooms, assuring good ventilation. Natural pine, stained dark, was used for woodwork, stark contrast against the white plaster that faced the coral rock walls. One of the interior walls was left unplastered, exposing the rock in all its rough beauty. "My skirts were always catching on the wall," mused Mrs. Gifford. Rock walls are common in modern interiors but were nothing short of daring in 1920.

Coconut Grove is studded with houses highly-prized by their owners which Dr. Gifford "had a hand in." Everyone knew everyone else in the Grove in the early days. If someone planned to build a house, they might informally ask Dr. Gifford what he would suggest.

Gifford gave freely of his advice and his time, concerned only that the house would be a comfortable dwelling for the tropics. One of the houses attributed to him is at 3564 St. Gaudens Road. It was built in 1919 of single thickness Dade County pine, naturally finished with creosote. Dr. and Mrs. Paul Robertson, owners of the house, recall that when they bought it Dr. Gifford walked over the property with them, identifying all its trees. "He was so immersed in trees that he thought in terms of them. He liked to use wood vertically, like it grows in the forest, on the theory that the rain should run down the grain of the wood. During hurricanes in this house the rain may stream down the boards, but it never drips in" says Dr. Robertson. The Robertsons entertain many foreign guests, and Mrs. Robertson says, "Whether people come from Thailand or Norway they invariably comment that something about the house reminds them of home. Dr. Gifford's ideas about shelter were so primitive, so basic, that they would fit in any part of the world."

The comfortable, unpretentious nature of Dr. Gifford's houses made little impression on the architectural trends of his times, but today his building style is coming into its own as reflected in some of the finest homes of Coconut Grove and Key Biscayne, and may some-day be recognized as the genesis of a regional architecture in southern Florida.

Civic Leader

Dr. Gifford was a member of the "Scientists Committee" of the Miami Community Council with Dr. Fairchild, Dr. Charles Torrey Simpson, Thomas Harris, and Edward Simmonds. Under the chairmanship of Frank Smathers, this distinguished group of botanists was appointed in 1922 to suggest trees for shading streets and parks in the raw, bulldozed landscape that was this burgeoning community of Miami. To the Miami City Council he advocated small municipal parks placed so that they would be seen by persons alighting from a train or traveling a busy street. Never one to hold back his feelings, he objected fiercely when Washington palms were brought in from California in 1920 to be planted on the new county causeway (now known as MacArthur Causeway). Dr. Gifford described it as a "desert palm from the West entirely unsuited to Miami's humid climate . . . It is a sacri-

The "Scientists Committee" of the Miami Community Council takes time out for a picnic lunch at Simpson Hammock during an exploratory tour around Miami in 1922. Left to right: *first person unidentified; Thomas Harris, English botanist from Jamaica responsible for the Deering plant introductions; Dr. Charles Torrey Simpson, then associated with the Smithsonian Institution; Dr. Gifford; Edward Simmonds; Dr. David Fairchild, head of the Bureau of Plant Introduction, U.S. Department of Agriculture.*

lege to plant it where the cocopalm or palmetto will grow. When it grows old, it wears Hawaiian skirts which harbor cockroaches and other small bugs."[40]

Mrs. Gifford remembers that though her husband had a warm-hearted, friendly personality, he could be very cold and formal when he felt that some politician was serving his own interest rather than that of the public. Gifford's views, strongly expressed on many occasions, won him few friends with the city council. Lewis Adams says that Dr. Gifford had his way of circumventing the council's decision to plant the Washington palm. He conspired with his friends to run over the young trees with their cars whenever they had the chance! This story is typical "Giffordiana" and it is impossible to tell whether it was true or just reflected his caustic way of speaking, but, like a parable, it shows how strongly he felt about planting the wrong tree.

He loved the Everglades. Perhaps their strange beauty recalled a boyhood memory of the vast New Jersey swamplands. By skiff he followed the waterways of the Glades into Whitewater Bay, through the great mangrove forest on the west coast and the Big Cypress Swamp. Once he was caught in the Everglades during a hurricane and sat the storm out in his flat-bottomed boat moored to a thicket of mangroves. His curiosity about Florida's early inhabitants, the Indians, led him to search the white beaches of Cape Sable for their mounds and kitchen middens. When, in 1921 the Florida Society of Natural History was formed to study ways to preserve South Florida's natural resources, Dr. Gifford was one of the charter members, along with R. D. Maxwell, Dr. David Fairchild, Gaines Wilson, Mary C. Lott, Harold B. Bailey, Colonel E. A. Waddell, Judge W. E. Walsh, Harold Matheson, and others interested in the preservation of South Florida's unique natural world.

Florida had no greater champion of her natural resources than John Gifford. Under "natural resources" he listed not only lumber, minerals, fruits, and vegetables but South Florida's coco palm, the mahogany, and the royal palm. He included the clam beds, the coral rock, roseate spoonbills, and wood ibis, in America native only to tropical Florida. He included such intangibles as geographical location and climate. Highest on his list of Florida's natural resources were its human resources, including the University of Miami. On one occasion he commented, "The unbeatable spirit of our people is one of our greatest resources. A man like Charles Torrey Simpson is worth millions to any country and can not be beaten by adversity even if a Miami policeman once tried to arrest him for vagrancy when he was returning from Key Largo at midnight with a sack of shells on his shoulder."[41]

A lifelong advocate of national parks, Dr. Gifford took every opportunity in speeches, radio broadcasts, and in his writings to urge the establishment of a national park in the Everglades. When Ernest F. Coe took the helm of the Everglades Park movement, forming the Tropical Everglades Park Assocation to bring the case for the park before Congress, Dr. Gifford was a staunch supporter. In 1931 Dr. Gifford was one of a group of Florida naturalists who accompanied Mr. Coe and some visiting scientists on an expedition to Cape Sable to demonstrate the unique and precious nature of the region. Many thousands of words were written and spoken by Dr. Gifford and other men

Seven-car motorcade about 1915 into an Everglades hammock

of vision before the Everglades National Park was finally dedicated in December 1947.

Roads were vital to southern Florida's development, and John Gifford advocated that they be built as quickly as possible, linking growing centers of population. His motion to continue "20th Street" (now Eighth Street) and the Orange Glade Road due west to Tampa is believed to be the oldest mention of the Tamiami Trail in print.[42] He urged the building of an overseas highway through the Keys to Key West in 1934. It was finally completed in 1938, accelerated, no doubt, by the destruction of the Florida East Coast Railroad route to Key West by the great hurricane of 1935. In the twenties he proposed linking Elliott Key and Islandia with the mainland. Fifty years later, the road and causeway are still being debated

It becomes apparent that Dr. Gifford was much more than a businessman, interested only in personal gain. Through every facet of his life, he shared his knowledge of the indigenous qualities of Florida, ever regarding scientific knowledge as in the public domain and giving of it freely. He worked with dedication for the responsible development and productive settlement of this rapidly growing American frontier.

Author—The Everglades and Southern Florida

Throughout his adult life Dr. Gifford was a prolific writer. He wrote for national publications long before coming to Florida. However, he became widely known for his articles about Florida life and tropical horticulture which appeared in magazines such as the *Country Gentleman, Wood Craft, National Geographic, Garden Magazine,* and other publications, including the little paper *Everglade Magazine* published by the Everglade Land Sales Company. "Plant your money in Dade County soil," its ads sloganeered.

A collection of Dr. Gifford's articles was published in book form in 1911 by the Everglade Land Sales Company. Entitled *The Everglades and Other Essays Relating to Southern Florida,* it sold for one dollar and by 1912 was in its second printing. Dedicated to the memory of Napoleon Bonaparte Broward,[43] it was written to "arouse interest in this great drainage project and to offer helpful suggestions to [Florida] newcomers."[44] The state had begun to drain the Everglades in

1906. The new land reclaimed for farming had increased the Everglades land owners from a handful in 1909 to 1,500 in 1911. Potato, tomato, and bean crops replaced the saw grass as drainage canals steadily lowered water levels. Dr. Gifford had visited the Landes of France and other land reclamation projects in the Netherlands and Denmark which were serving the needs of growing populations. Based on these observations he was heartily in favor of the Everglades drainage project in 1911. He later reversed his stand when the effect of the plan on the ecology became apparent.

In *The Everglades and Southern Florida* are articles on "sours and dillies" (limes and sapodillas) and the banana and the pawpaw. He recommended valuable trees for planting in the Everglades and described common and useful Florida plants such as the koonti, a primary source of starch. Many of his readers were farmers pioneering in the Everglades with no electricity, water, or sewage plants, to say nothing of modern housing projects! Dr. Gifford's practical handbook told them how to build water cisterns, electric wind generators, septic tanks, and comfortable bungalows designed for the climate. His articles were scholarly, but written in a most entertaining style, liberally larded with colorful language and anecdotes. About trees, he wrote as a forester of their usefulness as windbreaks, soil builders, and food producers rather than as ornamentals planted for decorative effect.

Planter-Developer: Elliott Key

Soon after settling in Coconut Grove, Dr. Gifford bought a large tract of land on Elliott Key, then known as Elliott's Key. This offshore island, the largest in the Islandia chain, could be reached by boat from Miami in about two hours. From New York, via the Florida East Coast Railroad, it was a forty-hour trip and offered a warm climate, boating, and the simple rugged life of primitive fishing camps. Many northern businessmen retreated to this pleasant tropical hideaway as an escape from winter weather.

Dr. Gifford published a pamphlet entitled "A Tropical Plantation Colony on Elliott's Key" in which he offered "twenty-acre lots facing both the bay and ocean and extending across the island with riparian rights for $1,000 each, only $50.00 per acre." He noted that lots had already been purchased by Dr. Charles Dolley of Philadelphia (his old botany professor), J. Horace McFarland of Harrisburg, and Dr. Charles DeGarmo of Ithaca, New York (former president of Swarthmore College). Many of Dr. Gifford's friends, probably through his encouragement, owned land on Elliott Key. Lewis Adams bought a piece with an old house on it and still owned half a mile of ocean front at the time of his death in 1968.

In 1969 the Islandia chain, including Elliott Key, was declared a national monument by the United States Congress. During the government's acquisition of this land in 1970, the United States paid owners as much as $7,500 per acre, and private land developers would have paid much more.

The ocean was the superhighway of the Florida pioneer. People traveled by boat to church, to work, to shop, to visit. "From some parts of the Keys," noted Dr. Gifford, "you had to sail forty miles just to mail a letter or buy a beefsteak." Dr. Gifford's boat was a seaworthy sloop named *The Dilly* that appealed to what he called the "buccanerish taint in my blood." Recalling his visits to his "farm" on Elliott Key, he wrote, "Three or four times a year when we want to bathe in the briny parti-colored waters of the Keys or seek plunder on beachcombing expeditions along the shores, I drop in to look over my plantation." On these expeditions he would be accompanied by his wife or a friend or two. No one wore special yachting clothing. He embarked in a soft, white linen suit, a shirt and tie, his head covered with a broad-brimmed hat. Edith wore full skirts to her ankles, a leghorn hat firmly tied on with a scarf, and gathered treasures from the beach in her skirts. Mooring the boat on the leeward side of the island where the breeze was heavy with the scent of the pineapple fields, Dr. Gifford would chat with Parson Jones, who cared for his grove. Speaking of the Parson, he wrote in *The Everglades and Southern Florida,* "The party who has charge of my lime grove is a colored man of more than ordinary intelligence. I have learned to listen to his statements. He lives practically alone among his trees. He is seldom bothered by the opinions of other men. His conclusions are his own. They are the product of the thorns and rocks with which he toils. 'Limes,' he says, 'and I guess other things too, must be planted close together so that the ground is soon covered. The lime is a half-wild crop anyway and the less you prune or meddle with it the better.' "

Gin rickeys had just been introduced in the northern cocktail

Parson Jones and his wife, Moselle, who tended Dr. Gifford's lime grove on Elliott Key.

lounges and limes were a good investment. The juicy, fragrant fruit netted Dr. Gifford about $3.50 a barrel. Prohibition in the twenties, unfortunately, virtually destroyed the market for limes.

Parson Jones, born Israel L. Jones, was to have a success story. An industrious and thrifty Negro, he came from the Carolinas in the 1890s when Dade County counted more Indians than Negroes. He tended several pieces of property in the Keys, including those of Dr. Gifford and Commodore Munroe. Munroe refers to him in his book, saying, "The clearing of the Cape in 1893 and the need of a caretaker for the property afterward, made the first step upward in the life of a young Negro, Israel Lafayette Jones, later well-known on the Bay as "Pahson Jones,' homesteader, fish-guide, and philosopher of Caesar's Creek. Black, strong and cheery, his ambition had brought him from the Carolinas to this land of opportunity when there were practically no colored men in Dade County, and after a period as a handy man at Peacock Inn, he married Moselle, an equally ambitious girl from Nassau, and they were established as guardians of the cottage on the Cape. Here their two sons were born, and Moselle appealed to Miss Mary Davis for suggestions of great men to name them for. Miss Davis could think of no more admirable heroes than King Arthur and Sir Lancelot, and so they were named. To Moselle the title was an integral part of the name, and they were always addressed by both together; I can still hear her musical hail from the doorway: 'Oh you, King Arthur! Come to dinnah! ' Jones roughed out most of the bush in preparing for the new house and grounds, and killed eight or ten rattlesnakes in the process, one of which was nine feet long."45

Parson Jones owned his home on Porkey Key and considerable property at the time of his death. His funeral was attended by many white friends as well as black and rated two columns in the *Miami Herald*. An item in the *Miami Herald* of January 17, 1970, noted that the sale of the Islandia Keys to the National Park Service is bringing sudden wealth to some early inhabitants of the islands. "The biggest purchase involved 182 acres on Totten Key sold to the Department by Sir Lancelot G. Jones, a bonefish guide who was born in the Keys. Jones and his sister-in-law, Kathleen Jones, received $800,000 for the Totten Key acreage and $55,000 for another twelve acre tract on Old Rhodes Key."

Boom and Bust in Miami—1920-1926

Between 1920 and 1926, Miami mushroomed from a pleasant city of 30,000 to a booming metropolis of 102,582. These were the golden days for John Gifford, both in terms of personal fulfillment and financial gain. By his own admission, Miami's boom time was one of the most exciting periods in his life. He bought and sold much land; won and lost a fortune. In Miami, he once held title to land from S.E. Sixth Street to the Miami River. Several of the apartment houses he constructed there still stand. He had real estate on Elliott Key and Miami Beach, including an apartment house strategically built at the spin-off of the dog-legged MacArthur Causeway. With a partner he bought Brickell Point in Miami for $1,500 and later sold it because no one could see a future in that piece of muck, marl, and mangroves. Today the land is worth several million dollars.

Personally he was saddened by the weakening health of his wife, Edith. Her condition became so critical that she was taken to a hospital in the North for special treatment. There she died of pernicious anemia. Dr. Gifford's mother came to live with him in Miami and "End of the Trail," the gracious, red-tile roofed house they had built on the bay, was sold. Dr. Gifford converted the former packing house of his grapefruit grove into a home, the nucleus of the handsome rock and pine house on S.W. 27th Avenue.

By 1920 the MacArthur Causeway was opened, a two-lane highway skipping across Biscayne Bay to connect Miami and Miami Beach. As permanent residents, lured by nationwide advertising and promotion, streamed into America's newly discovered tropical paradise, the real estate market soared. Responsible developers like Dr. Gifford who were concerned for the long-range growth of the area feared for the future as "get-rich quick" salesmen invaded the market. Called the "binder-boys," they would place a small down payment on a piece of property, cut it up into subdivisions, and sell it at a profit, all in one day. Profits were mostly on paper, but many men became millionaires overnight. Land close to Miami that Dr. Gifford had sold for $200 to $500 an acre in 1924 was selling for two or three thousand dollars an acre in 1925. When Dr. Fairchild wrote to Dr. Gifford asking his advice on investing in real estate, Dr. Gifford wrote, "The price of land here has gone beyond all reason. I would fear it were it not for the fact that men with money are buying. I can not see how the bubble can burst if there is real money behind it and that seems to be the case."

Some 7,500 real estate licenses were issued in 1925 alone, and 174,530 deeds and papers were filed by the county clerk. The cars of the Florida East Coast Railway were backed up for miles with lumber and building supplies waiting to be unloaded. In the harbor at the foot of Flagler Street a forest of three-masted schooners carrying lumber jammed the wharfs. Two hundred, eighty-one hotels and apartments were built in a twelve-month period as well as uncounted private homes.[46]

The shadow of depression was deepening all over the nation and by the close of 1925 the Federal Reserve Bank had begun to stiffen rediscount rates and curtail loans. The word in northern financial circles was that Florida banks were risky and some real estate operators began to face up to heavy income tax on profits still on paper.

In 1923 Dr. Gifford married Martha Wilson, her second marriage as well, and she brought with her a little daughter, Jeannette, now Mrs. Thomas Smith. Dr. Gifford's first marriage had been childless, to his great regret, and he was delighted with the child. The Giffords later had two daughters of their own, Emily Jane (Mrs. Cotton Howell) and Martha (Mrs. Harold Melahn), as well as fourteen grandchildren.

Author: Billy Bowlegs and the Seminole War

Billy Bowlegs and the Seminole War was written during the boom years. It was published in 1925 by the Triangle Company in Coconut Grove and printed by Lewis Adams' Kingsport Press. A strange little volume, *Billy Bowlegs* was not quite a book at all, but rather a long essay inspired by an article entitled "Billy Bowlegs in New Orleans" which appeared in *Harper's Weekly* (June 12, 1858). Just seventy-nine pages long, the volume was well-illustrated with eleven plates, several of them fine drawings of Billy Bowlegs, chief of the Seminoles, his young wife, and Ben Bruno, a Negro slave. Dr. Gifford used the *Harper's* article as a vehicle for his observations about the Seminoles. He had visited their villages in the Everglades, bringing gifts of cloth and tobacco, the latter of which brought the women scurrying out of the brush faster than the cloth! Though he recognized and freely admitted the weak points of the Indians, he appreciated their simple dignity and regretted the encroachment of the white man's civilization on their simple life. Like Thomas Barbour, he suggested that the Indians be trained as guardians of the Everglades National Park.

Mary Street from the Bay after the 1926 hurricane

Reviewing *Billy Bowlegs* for the *Miami Tribune* (June 4, 1925) Charles DeGarmo wrote, "We who have had almost daily sight of Seminole men, women, and children as they wander through our streets or camp by our roads, cannot fail to be interested in Dr. Gifford's sympathetic revelation concerning Seminole habitations, intelligence, views of life, conduct, personality and appearance. He confirms many of our surmises, corrects our erroneous opinions, and gives us many a gleam from an unexplored mental and emotional cavern."

Big Wind of 'Twenty-Six and a Fortune Lost

In September 1926 Miami's weather, usually her greatest asset, delivered the final blow to the land boom. The front page of the *Miami Metropolis* on September 17 carried a streamer warning of a tropical storm reported blowing one hundred miles an hour and heading toward the Florida coast. Ship owners made their vessels fast in the harbor and small craft were moved into the Miami River. Few of Miami's settlers, more than fifty percent "greenhorns" to the tropics, had any experience with a hurricane. They knew all about blizzards and tornadoes, and expected little more from a hurricane. The wind began to blow late that night. By the following afternoon it was all over, and so was the phenomenon of Florida's first land boom.[47]

The nation's papers carried the headlines, "South Florida wiped out in storm." Red Cross relief trains sent from New York could not even reach the area for three days because the tracks were strewn with debris. One hundred and fourteen persons were dead, 25,000 were homeless. In the Miami River and Biscayne Bay nearly one hundred

fifty vessels lay on their sides, severely damaged or sunk and blocking the harbor entrance. Kenneth Ballinger in his book, *Miami Millions,* called it "the greatest catastrophe in the history of the United States since the earthquake and fire in San Francisco." *The Coconut Grove Times,* its printing plant wrecked, published a four-page mimeograph edition on September 24 describing the damage of the hurricane in the Grove. Mercifully and miraculously as well, only one life had been lost, though many injuries and narrow escapes were reported. Homes were blown down or battered, and some were roofless. Boats were blown inland, resting high up on MacFarlane Road. One large boat smashed and demolished the William Catlow cottage on the bay and the family had to seek refuge with neighbors. Then rising water forced them into the attic where they sat out the storm. Commodore Munroe's boathouse where he kept the plans and drawings of a lifetime's work as a ship architect was swept away and all its contents were lost. A Mr. Traphagen, the news sheet reported, clung to a chandelier in seven feet of water for four hours after the grand piano on which he had been standing was swirled away by the invading sea water. A twelve-foot beam charged through the house of Mr. and Mrs. H. DeB. Justison and lodged on the kitchen sink without damaging anything en route. Many sailors tried to ride out the storm on their boats, but most were found clinging to a treetop or the hull of their capsized boat. Dinner Key was swept clean of boats and dwellings. Trees were down and stripped of leaves and most of their branches. Among the fallen were Dr. Gifford's two original lofty fig trees on St. Gaudens Road that had been the object of many a tourist's lens.

The same circumstances of an uninformed populace, combined with lax or nonexistent building codes and a harbor bristling with the masts of hundreds of vessels loaded with lumber and other building

U.S. Plant Introduction Garden at Miami shows ficus trees uprooted by the hurricane of September 1926.

supplies, never combined again in Miami. Though hurricanes today may damage plant life, buildings, and electrical installations, it is doubtful they ever again will cause such tragic loss of life and property.

John Gifford and Lewis Adams weathered the storm safely in the house on S.W. 27th Avenue which Gifford had just expanded again for his growing family. The roof lost hardly a tile. John Gifford knew how to prepare for a hurricane and his houses held fast. Not so his fortune. His formula for financial security—investments in small home mortgages, land, and bank stocks—proved untrustworthy. With the collapse of the southern Florida real estate market and the nationwide depression, mortgages were worth little and the banks failed. To compound the problem, Miami city officials were reluctant to admit the collapse in land values, and continued to assess land at boomtime values. Paying taxes on his extensive land holdings further dissipated his wealth. "Pity the man who has to sell his wife's jewels to pay his taxes," Dr. Gifford later wrote.

Asked how Dr. Gifford took his financial reverse, Lewis Adams replied, "He sprang back like a cushion with springs in it! Over the next few years his income shifted from $300 a day to the twelve hundred a year he made teaching forestry at the University of Miami. Looking back on it later he told me that it was probably the best thing that ever happened to him."

Henry Troetschel, Jr., in an article about Gifford in *Tequesta,* [48] says, "The collapse of the Morris Plan Bank was a source of great sorrow to Dr. Gifford. In later years he spoke but little of the incident and once mentioned that this had been his biggest failure.

Dean Russell Rasco of the University of Miami Law School tells that he had done legal work for Dr. Gifford in connection with the collapse and that they always had been warm friends. One day, coming across the dean in his office, Dr. Gifford had said, 'How's the old shyster these days? ' Rasco, answering in the same vein, had inquired, 'How's the old broken-down banker? ' Dr. Gifford became very angry about the exchange. His reaction reached the dean in a letter in which Dr. Gifford made it clear that he felt Rasco had not spoken with proper respect. Rasco telephoned him immediately and smoothed his feelings."

Dr. Gifford was one of the organizers of the Florida Exchange Bank, which opened in 1926 on Old Main Highway in the business district of Coconut Grove. It merged with the Coconut Grove Bank and Trust across the street in 1930 and became the present Coconut Grove Bank. Dr. Gifford served for the rest of his life as a member of the board of the Coconut Grove Bank. A. B. Harrison, president of that bank in 1970, served with Dr. Gifford on the board of the bank as a young man. He recalled: "His presence on the board meant so much to us when times were tough. He had vision and believed in the growth of this area. What's more, he could communicate it to others. People had confidence in him. He never knew a stranger and was a friend to everyone whether they had a million or not a nickel." W. T. Price, chairman of the board of the Coconut Grove Bank in 1970, explained that Dr. Gifford was not a speculator during the land boom. "He bought land and sold it at a profit in a responsible way. When the boom collapsed we lost a lot of money, but neither of us went into bankruptcy. There were darn few that were doing any business that didn't," said Mr. Price.

Dr. Gifford while a professor at the University of Miami

University Professor

At heart, John Gifford had never stopped being a teacher. When in 1931 the opportunity arose to teach an evening course in tropical forestry for the downtown branch of the University of Miami, he accepted. His students were mainly adults, nature lovers, landscape specialists, home gardeners. With his vigorous enthusiasm and dry humor he charmed them all. One of his favorite ideas was to plant mango trees along the curbs of the city's streets. Pointing out that other tropical countries used mango trees for shade and beauty, he insisted they would be an asset to Miami's residential streets. "But," he complained, "the authorities don't like 'em. The tree has a grievous fault; it is a social offender. It bears fruit! This is its great sin, dumping this fine food all over the ground and inviting small boys into its friendly branches."[49]

The University of Miami was still in its formative stages. It had been chartered in 1925 just before the depression hit southern Florida, and its development had been delayed by the financial crisis. Its founders, however, had high hopes. From the beginning they were aware of the university's unique geographical location. The university charter speaks of its intention to form a "program dealing with the scientific and technical problems relating to the tropics," and all botany department courses beyond the junior level emphasized tropical and subtropical botany.

Dr. Gifford was made a full Professor of Tropical Forestry in the Botany Department in 1933, a post which he held until his death in 1949. He taught General Forestry, Forest Policy and Tropical Forest Industries, and Geography 131 (Conservation of Natural Resources). Geography 131 was required for certification of public school teachers and with these students he made a lasting influence on his state. In his declining years literally hundreds of these young people called on him at home. Sometimes they were servicemen returned from the Pacific in the Second World War. What they had learned in their brief course in "tropical forestry" with old "Doc" Gifford had helped them survive in the jungles. His classes were among the most popular on the campus, always over-registered, and even packed on Saturday mornings! He was a campus institution. One year the whole football team took his course. If attention lagged he would invite the class out onto the campus and, with a blowtorch, demonstrate the fireproof qualities of the cajeput. Some said that he never gave a failing mark, that his course was a "snap," and it may have been, but few students left his room without a deep appreciation of the natural things of John Gifford's Florida. "He was really a philospher," explained Dr. Alexander, "and he had a message. I think he was trying to say, if you don't go along with nature's laws, you are going to have trouble."

Like all great teachers, Dr. Gifford spoke in parables, telling many

Notes for a class lecture. From Dr. Gifford's notebook used when he was a professor at the University of Miami.

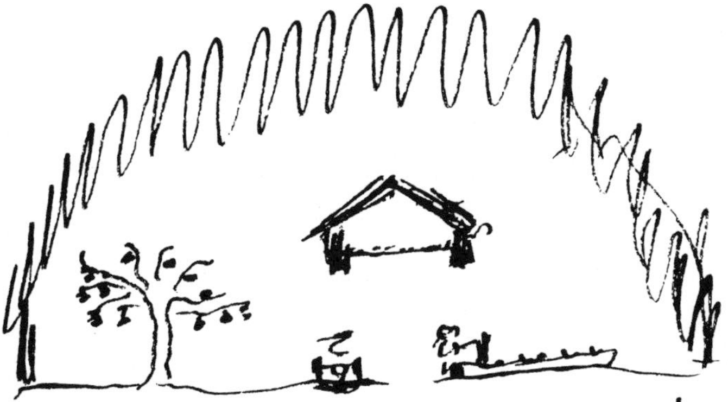

Homo sapiens Sylvanus
Homo sapiens - campestrian
Homo sapiens - urbanus

campestris

House plan   of the
                fields
nat. Yurt plan   or
                plains

Five, Wonders of
       Florida.
Jumon | Torreya
Silver Spring
Royal Palm Hammock
Clambank -
Cypress Sovereign.

a story along the way. His late son-in-law, Henry Troetschel, describes a typical lecture. "His rambling may cover half a continent, criss-crossing every which way, turn around and come back again, but eventually it reaches the sea. Asked why he didn't make outlines he answered, 'I know what I want to say. I've said it enough before. And I tell you, you just can't get most people to understand what you're talking about unless you tell them again and again . . . and lots of people will enjoy it if you tell them some stories along the way and just keep slipping in what you really want to say.' "

The students loved Dr. Gifford's lectures because they were full of humor. If a story didn't have enough humor in it, he added something. Lewis Adams, visiting his friend's class one day, heard him telling about a "little runt of a taxi driver" provoking a big husky bus driver into a fight. The bus driver was in the right and, after he had taken much abuse from the little guy, gave him a good sock in the jaw, Dr. Gifford told the class with great relish! Mr. Adams had been there when the incident occurred, and the only sock in the jaw was the one Dr. Gifford and Lewis Adams had *wished* had been given! When Dr. Gifford spotted his old friend at the back of the room he knew he had been caught embroidering a tale. Adams quickly joined the conspiracy and confirmed his story. "He often embellished the truth to make it more humorous, but he never told a harmful lie. In fact he was reluctant to speak unkindly about anyone," recalled Mr. Adams.

Dr. Gifford taught in the old "cardboard college," so-called because the rooms were divided by nothing but thin pressboard. Classes were held in the Anastasia Building in Coral Gables (since demolished), a makeshift measure until the college could get on its feet.

Dr. J. Riis Owre, Professor of Spanish, came to the university in 1935 and remembers Dr. Gifford well. "I was a newcomer to Florida and called on Dr. Gifford one day to ask his advice on what to plant. The old Gifford place was a regular jungle. He told me, 'Whatever you plant, don't fertilize, water, or weed it. Leave it alone. If it wants to grow it will. If it dies, it should not grow here.' He often said he didn't believe in sitting up nights with plants."

His voice was strong and carried far. Some faculty members objected at being placed too close to his room. Once he was heard expounding, "Cows don't sweat, I tell you. Cows don't sweat! " Dr.

Gifford took some good-natured kidding on this score as the word was solemnly passed around the faculty, "Cows don't sweat, you know."

Dr. Gifford wasn't the only one who came in for a ribbing. There was a good spirit around the university. It was a small faculty, no more than fifty, and the student body didn't swell above six hundred until after the Second World War. Teaching conditions were far from ideal with everything from instrumental music to conservation being taught in one building, the rooms facing each other across an open courtyard, but there was a strong feeling of comradery. Looking back on those early days, Dr. Owre remembers, "We all felt we were in on the beginning of something good."

Dr. Tebeau remembers that booming voice too. "All the students within its range took his course! " One student commented to Dr. Gifford how much he had enjoyed his course. Dr. Gifford couldn't recall having him in his class and the student explained that he was in the room across the hall.

"He took a lot of kidding," Dr. Alexander agrees, "but he didn't get mad about it. This was a small group of professors who knew each other well and had a common goal: that of building a university."

Dr. Leonard Muller, Professor of French, took two of Dr. Gifford's courses in 1936 while earning his American degree. When he wanted to sign up for another Gifford course, the older professor advised against it. "It's the same course," he said, "I tell the same stories." In spite of the informality of his courses, much was learned. Dr. Muller admitted that little was expected in the way of outside class preparation, but said that what he taught was long remembered, like: "Respect the common things. They are common because they have survived," or "A weed is just a plant out of place."

Highpoint of Dr. Gifford's course would be a field trip in a rattling old Ford truck from Miami over the Keys to Key West with a possible side trip by boat to Elliott Key. Along the way, Dr. Gifford would point out hundreds of plants and trees, giving their scientific and common names and telling all the stories he knew about them.

Author: Three Books in the Thirties

In the thirties, while teaching at the University of Miami, Dr. Gifford wrote three books: *The Tropical Subsistence Homestead* (1934),

The Rehabilitation of the Floridan Keys (1934), and *The Reclamation of the Everglades with Trees* (1935). Bound together they scarcely amounted to three hundred pages no bigger than a hand, but they contained Dr. Gifford's most cherished ideas for the development of the Keys, southern Florida, and the Everglades. They were published by Books, Inc., printed by Lewis Adams' Colonial Press, and distributed through the University of Miami bookstore.

The Tropical Subsistence Homestead

This theme is interwoven in all Dr. Gifford's teachings. It stems from his conviction that the closer a man is to the earth, the better person he is. He considered the small farm home the essential basic unit of society. "The prosperity and strength of any country can be measured by the number of small, self-supportive homesteads which it contains," he wrote.[50]

In his book he offered the subsistence homestead as the answer to the nation's problems of unemployment and breadlines. Not only poor white people, but Negroes and Seminole Indians could be comfortably settled on five-acre plots which would be offered for sale at reasonable prices and low interest rates. Though in later years he recognized the problems of overpopulation and the rising cost of land, he clung to the idea that the country still contains much land regarded by many as "marginal" which can be productively cultivated with the right tree crops. "Often it is the man, not the land that is marginal," he said.[51]

Dr. Gifford found little support for his return-to-the-land schemes, but he held fast to the tropical subsistence farmstead concept throughout his life.

"This was Dr. Gifford's projection of the Jeffersonian dream of the yeoman farmer," commented Dr. Tebeau about the five-acre homestead idea. "It is not important today economically. It is important socially."

Dr. Alexander, too, agreed that men are always better off if they are capable of caring for their own needs. However, he could not see a return to the days when man was self-sufficient on his own land. "Dr. Gifford never saw the invasion of the Mediterranean fruit fly or anticipated the impact of population pressures. His times did not have this kind of complexity."

The Rehabilitation of the Floridan Keys

The depression hit southern Florida particularly hard, especially Key West. The naval base at Key West had been closed, and competitive markets had cut into some of the industries on which this city, once the largest in Florida, had depended. *Rehabilitation of the Floridan Keys,* dedicated to the memory of Dr. Henry Perrine, the pioneer plant introducer, was a comprehensive plan for restoring economic health to this colorful region. In it he wrote of the natural beauty of the sea and shore as the Keys' greatest resource and urged their preservation. He advocated the establishment of the Everglades National Park to include Fort Jefferson in the Dry Tortugas.

He wrote of Dinner Key, then an international airport whose giant "flying clippers" winged over Dr. Gifford's house just a few short blocks from the landing field. Dr. Gifford saw Dinner Key as the hub of communication between North and South America and proposed air tours for tourists to the national parks, starting with the Everglades National Park and extending to the Luquillo Forest Reservation and throughout the islands of the Caribbean. Of course, the parks were still to become a reality, but John Gifford had confidence in the future.

He envisioned another travel route, "the coral trail," a system of highways and ferries crossing the reefs and stepping over the Keys between southern Florida and the countries of the Caribbean. The coral trail would transport goods, people, and customs from Key West by ferry across the ninety miles of water to Cuba, east to Puerto Rico, west to Cozumel, the Island of Swallows, and finally to the Yucatán Peninsula. The coral trail is well-traveled today, but through the air and by rent-a-car systems. It remains for some future engineering expert to make possible Dr. Gifford's dream of a highway across the water.

Always looking into the future, Dr. Gifford anticipated many modern trends. "Rich men will buy islands, and build winter homes there," he wrote of the West Indies.[52] Gardens on the sea bottom will be cultivated for sea truck,"[53] he predicted, envisioning the day when sea farming would be big business.

Reclamation of the Everglades with Trees

The third little book was *Reclamation of the Everglades with Trees.* He once said, "I have no hesitation to cut a tree to substitute in its place a better kind. I have no hesitation to junk any idea for something that is better."[54] This book represented a reversal of his ideas about draining the Everglades as expressed in his earlier writings.[55] Once a great supporter of the idea, he admitted its shortcomings as the work progressed. The black muck of the Everglades which, when first drained and planted to crops produced fantastic yields, proved too shallow to support plant life over a period of years. When the engineers broke through the rocky rim of the natural cup-like dam that contained the water in the Glades, the water drained out all right, but the salt water of the bay flowed in as well, ruining crops. Fires, once confined by water breaks, swept over broad areas unchecked, threatening to make a Sahara out of this unique region. Severe flooding during the hurricanes of 1926 and 1935 resulted in tragic loss of life and property, and further discouraged the plan to convert the Everglades into farmland. In *Reclamation* Dr. Gifford suggested that when population needs demand reclamation of this area, it might be done with trees, which he called "nature's own drainage pumps."

While presenting his idea, he wrote about a hundred other things. The reader learns about food medicines such as the elderberry; a grass (bamboo) that is classed as a tree; a tree (coconut) that is a "supermarket"; and his philosophy about pencils, cigar boxes, and the octopuslike habits of the fig tree.

Father

During his later years he spent much time at home and took pleasure in his growing family of girls. Though he was past fifty when his first child was born, fatherhood was natural to him. "He always loved children," says Lewis Adams, recalling that they always gravitated to him. "They knew they had found a buddy," was how he put it.

His daughter Jane says, "We never had expensive toys, but we had something better—a father who was devoted to us. I was his pet and often played under his desk when he was writing. He made great things for us and everything was consistent with his philosophy of conservation. Once he gave us an old water cistern and a pole to roll it around the yard. When it rusted out, he poked a pole through the hole, tied a flag on it, and it was a fort. We didn't have to worry about keeping off the grass either. Dad considered lawns a "northern tradi-

Dr. Gifford with daughter Martha; note the Indian rug

tion" totally unsuited to the tropics. We had every kind of plant but grass! Once or twice a week he would pile all the kids in the neighborhood into his old Ford and take us down to Tahiti Beach. He never went in the water himself, but sat there on the sand, his eyes shaded by an old felt hat, pipe in hand, watching us. He adored his grandchildren and they all teethed on his pipe stems." Daughter Martha told of the marvelous "trains" he would make for the grandchildren out of three or four packing boxes, and "volcanoes" that really smoked and exploded!

Always kind and indulgent with children, Jane says, "I think he got worse as he grew older." A visitor while talking to Dr. Gifford was distressed to see one of the children banging a spoon on the carving of an antique chair. When he drew attention to this sacrilege Dr. Gifford just gazed blandly at the offender and said, "Yes, they like to do that."

Jeanette says, "The only time I ever saw him really mad was when one of the cats had a litter of kittens on his clean, white shirts!"

Author: Living by the Land

Dr. Gifford's best known book, *Living by the Land,* was published by Glade House of Miami in 1945. Written when he was seventy-five, it contained a lifetime of thoughts about conservation and the right use of land with special reference to Florida and the Caribbean region. Marjory Stoneman Douglas in her review of the book for the *Miami Herald* on January 27, 1946, said: "Everything he has to say is important to those people now flooding in here. The people who buy lots and put bulldozers to the work of destroying every living thing on the ground, destroying our native hardwood, rooting up trees, often rare and of the greatest possible interest, should be required to buy and read this book. Nothing could have been more important or timely than that Glade House should have printed this book just now when we all need to be reminded again of the great facts of conservation and tree planting which he has tried with infinite patience year after year to teach us."

The Last Years

Though a world traveler, as Dr. Gifford grew older he preferred his own backyard to any other spot. In fact he was writing another book which was to be called "Backyard Browsing." The five-foot library table which served as his desk became his ship, and the twenty species of trees he could see from that window were the only ticket he needed to faraway places. Each tree represented a long chain of events and people, beginning with the person who developed its seed, refining it into a useful tree fitted to its environment. Letters, pamphlets, newspapers, magazines, and books cluttered his desk's surface and reflected his broad interests. Grace Norman Tuttle in her column "Echoes of Miami" (*Miami Herald,* July 7, 1931) tells about the range of topics a visit with the old forester might include. "Flying machines, Calusa and Seminole Indians, Elliott's Key, and what to plant in the tropics; all variations of his one great theme, Florida."

Evenings, Dr. Gifford held open house for the many visitors that pulled the old brass bell on the porch. If Lewis Adams was in town, he

Dr. Gifford in a favorite rocking chair

was likely to be one of the callers. Mrs. Maude Smith of Key Biscayne, Adams' daughter, remembers visiting the Giffords with her family as a little girl. "All the children liked to hear him talk," she said. "It was always so humorous, but the humor was attached to something of importance. He loved his home, but he was always willing to go." And where did they go? "Oh to see a tree up in Stuart [Florida] or some new development in town; an ice cream factory, for example."

Often Lewis Adams and Dr. Gifford would spend the evening smoking their corn cob pipes (they had had to give up the fancy Turkish tobaccos of their youth for the milder Sir Walter Raleigh) and philosophizing about life, man, and destiny. Both had been business-men, experienced wealth, a depression, and two world wars. "We were agreed on one thing," said Lewis Adams. "Man is at his best when he is carrying a heavy load uphill. When he sets it down, look out!"

Dr. Gifford's mind was a storehouse of information and strange tales, especially those related to trees. He could talk about oak casks, and though a temperate man, the wine that was in them. He spoke of buttons made of blood and plants that eat flesh. Even his government reports were entertaining reading.

Given to warm friendships and open-hearted hospitality, in many respects he remained solitary. He often asserted that trees should be planted in forest formation, close together for protection. He claimed no such protection for himself. A member of the Swarthmore Meeting House throughout his life, he never joined a congregation in Miami. Politically a Democrat, verging on a socialist in that he believed that resources such as water and electricity were the natural right of all men, he never ran for office or took formal part in a political campaign. He seldom stood before an audience without suggesting some project for the common good—roads, public parks, preservation of the sea gardens, a public museum of Florida artifacts, for example. His influence was widely felt and many of his ideas adopted, yet he led no movements and was never a manipulator. He preferred the role of a catalyst.

When the Fairchild Tropical Garden[56] was formed in 1938 honoring his friend, David Fairchild, Dr. Gifford's name did not appear in the dedication program. It was not surprising. When the ribbons were cut and the speeches made he was more likely to be at home writing up some idea he had had the night before. Then too, by his own admission, he was a "forester, interested in forests, not in single trees. I have no interest in collections of palms or collections of any one tree."[75] He did, however, serve on a committee to advise on plantings at the garden.

Never a joiner, Dr. Gifford nevertheless belonged to many botanical and historical societies. He was a life member of the Florida State Horticulture Society, a former president of the Historical Association of Southern Florida and of the Florida Botanical Garden and Arboretum Association, and an elected fellow of the American Association for Advancement of Science.

In 1932 he was elected an honorary member of the American Forestry Association "in recognition of your long and constructive service and leadership in forestry to which you have made contributions of deep influence upon both the profession and upon the men

Dr. Gifford examines a strangler fig feeding on a palm.

who have followed along the trails blazed by you." Ten years later he was elected a vice president of that organization to honor his work in conservation.

Many public tributes were given him in his lifetime, but his widow, Mrs. Martha Gifford, says that the honor that pleased him most was his election in 1942 as a fellow in the Society of American Foresters.

"He was deeply touched by this recognition by his colleagues in forestry," she said.

He attended few meetings except as a speaker, preferring to read about the proceedings from journals and publications. He was a most gregarious man, yet, paradoxically, part of him lay submerged in his thoughts about the forest, its creatures, and their relationship with man.

Aside from some intestinal troubles which he controlled with a self-prescribed diet, Dr. Gifford was generally robust. James Hodges, in a sketch of Dr. Gifford (*Miami Herald,* September 8, 1940) wrote: "At seventy, Dr. Gifford is sturdy and healthy. A fringe of white hair surrounds his bald head. Blue eyes look intently through his rimmed spectacles. His fingers, big and strong, break matches in two after lighting his pipe. This is a habit gathered in the forest to keep from starting forest fires. Looking at the pile of matches he has struck to keep his pipe alight he jokes, 'I have burnt up enough wood to start a forest.' "

In 1947, while lecturing at the university, Dr. Gifford fell from a lecture platform and broke his hip. Though the bone healed completely, he used the accident as an excuse to limit his public speaking to his university classes. Two years later he contracted uremic poisoning brought on by neglect of a prostate gland condition. Though an operation could have been performed, he preferred to treat the ailment himself.

Unfortunately, this was one thing his natural remedies could not heal, and he soon had to be hospitalized. From his bedside at Jackson Memorial Hospital in Miami Mrs. Gifford replied to a note she had received from Dr. Fairchild. "Dear Dr. Fairchild," she wrote. "It was my pleasure to read your last two letters to 'Dr. John' as he sat in his chair. He was responsive to the warm friendship and memories recalled through shared interests. You will be glad to know this. As to your inquiry, Dr. Gifford said he was at Tomellin, a town in Mexico in 1911. The monastery there gave him the fruit. There was no hesitancy in his memory of time and place. I wish that I could speak positively about his condition. We live one day at a time, in a sense that only a critical illness makes possible. We hope to have Dr. Gifford at home on Monday. He has been here four weeks. Speaking for the girls and myself, we are deeply grateful for yours and Mrs. Fairchild's expressed sympathy. It seems so real in a world of frailty and frustration. Sincerely, Martha Gifford."

Dr. Gifford died several days later on June 27, 1949, at the age of seventy-nine. His body was cremated and funeral services were held at the Coral Gables Congregational Church. Among his honorary pall bearers were Lewis Adams, A. H. Andrews, Dr. Bowman F. Ashe, first president of the University of Miami, Dr. Fairchild, and a score of other friends.

The *Miami Herald,* whose pages had carried many stories by and about this colorful figure, noted, "Not only this community, but all of Florida as well, has lost one of its most distinguished citizens in the death of Dr. John C. Gifford . . . He was a giant of a man. The years made no appreciable inroads on his seemingly inexhaustible store of energy. His booming voice never ceased to tell the story of the forests, man's abuse of them, and the need for conserving the trees that a bountiful nature has given us. He has left a notable impression on its [Florida's] structure and its progress."

American Forests (August 1949), the offspring of Dr. Gifford's original *New Jersey Forester,* spoke of his passing, saying, "The nation lost an eminent citizen and the world one of its most distinguished foresters in the death on June 27 of Dr. John C. Gifford, internationally known forester, scholar and scientist. Closely associated at the turn of the century with Dr. B. E. Fernow, Dr. J. T. Rothrock, Gifford Pinchot and other far-sighted men in pioneering the forest conservation movement in this country, Dr. Gifford has left a notable impression on its progress."

Few memorials remain of this uncommon man save the marks he made on men and on landscapes—the cajeput, the lofty fig tree, the coral tree from Jamaica, and a tree hibiscus from the West Indies, to mention some of the trees he is credited with introducing. In Coconut Grove, a street, Gifford Lane, just three blocks long, bears his name. He once lived there in a comfortable frame house he had formed by joining together two small pine cottages. At the University of Miami, a Gifford Society was formed by botany students, but it survived only a few semesters. The Gifford Arboretum at the university honors him as well. Planned and planted by the Botany Department in 1948 and 1949, it originally contained specimens of six hundred native and introduced trees. In recent years the Arboretum has deteriorated through hurricane damage and lack of funds for proper maintenance. Vandals have removed the identification plaques from their pedestals. Some trees have died and others need pruning and cultivation. The land itself is often eyed by the university for a parking lot.

Surely the greatest legacy of this early environmentalist were the ideas which he communicated throughout his lifetime. Though few were original with him (he never claimed that they were), still, he had a way of presenting them that made them stick in the memory. Those ideas come alive again, burning bright and clear from the pages of his little books.

One of his chapters in *Living by the Land* is called "Twig Culture." In times of colossal waste, it is refreshing to be reminded that there are those who value twigs and sticks and that the United States dime bears a fasces, or bundle of sticks, once the symbol of unity and strength. If John Gifford had believed in sin, the greatest would have been waste. Even a hole was valuable to him. "Look on a hole as an asset rather than a liability, a valuable aid in growing plants, not just an unsightly nuisance," he wrote in a newspaper article.

People have become so estranged from nature today that they cut down trees because they are "messy," but not a leaf or blade of grass was ever wasted on John Gifford's property. All were turned back to the soil. No high interest rates or installment plans for this man of independence. "A high standard of living is desirable, but there must be a limit. It is irksome to be too closely budgeted and to pay installments on things long after they are ready for the junk pile."58 What apartment dweller, dependent for his life on services of others, would not "yearn for a little complete self-supportive homesite to crawl into like a hedgehog in case of need."59

His friends shake their heads when questioned about the idea to which he devoted his life; small, independent tree farms where "pigeon-breasted machine workers" would be restored to manhood, walking tall and free in the land, supporting their families from the hand of nature. Yet John Gifford was a man gifted to see down through the centuries, beyond his nation, not yet two hundred years old, and his civilization. In an address to the Florida State Horticultural Society he wrote, "When man devastates the earth he lets loose the furies which are his own undoing. There are countless buried civilizations covered with jungle. In time the jungle always wins. Time is nothing to nature, but much to man. Whenever the works of man are contrary to natural law they vanish in time."[60]

And so in 1945, with dogged determination, he wrote: "Movement back to the land is altogether a practical scheme since, despite our growth in population, we still have sufficient space for the purpose."

As the urban crisis grows in America, social scientists are discovering, almost with surprise, that vast stretches of the land lie empty. Eighty percent of the people live in five percent of the land, it has been estimated. National attention is being focused on developing means of channeling the growing population away from overcrowded cities into the countryside. While the details of Dr. Gifford's independent homesteads need translation into modern terms, perhaps something can be learned from the rereading of the philosophies of this veteran forester and conservationist who believed strongly enough in his idea to hold it throughout his lifetime.

Part II

ON PRESERVING
TROPICAL FLORIDA

Excerpts from the Writings of
John Clayton Gifford

1 The Meaning of Conservation

Man is merely a custodian of many things and when he passes on he leaves everything behind, even the ruins of what he destroyed.

Living by the Land, p. 12

Truly, there is little that is new about conservation. Nations of the past, when their resources began to dwindle, hurried to conserve. Some were too late. A reflective analysis of history shows that all peoples, even though widely separated by time and space, go through the same stages of development. The story of civilization normally repeats itself, over and over again. Nevertheless, for all our wisdom, we fail to learn from archeology and history. We sagely criticize our fathers for lack of foresight—but fifty years from now our children and their children may look upon what we are doing today as incredible and foolish. We roll down the ages impelled by agonizing needs, blundering along because there are many truths that have not yet been discovered, many inventions and techniques that have been pigeon-holed, and many people who lack vision.

Do not think that conservation is merely saving and hoarding things. Conservation means *sane use.* Some nations began with nothing except mediocre land to cultivate and a sea to fish in, but they

(opposite) *Dr. Gifford with live oak in Matheson Hammock*

achieved a high stage of civilization, comfort, and self-sufficiency. Others, endowed with abundant natural resources, wasted or never had sense enough to use them properly; consequently, they failed. What is wasted today may be useful tomorrow.

How real is progress? The mighty machine that presses out a steel rod is not very different from the hammer we use to drive a nail. The complicated lathe is only a modification of the common wheel. The powerful dipper dredge is like the razor-back hog. The suspension bridge is like a spider web. We gain time and we defeat distance by these inventions, but machines do not bring unmixed blessings. Used unwisely, machines may destroy us. If we use an adding machine continually we may forget how to do simple arithmetic. "Conservation," Theodore Roosevelt said, "is the preservation of things by wise use."

Conservation is a kind of philosophy of living, worked out to some degree ages ago under various names. It is not so much a study of any thing in itself as it is a study of man's relation to things. Conservation is, therefore, a branch of ecology. This relationship between man and his environment can be disjunctive or conjunctive, reciprocal or antagonistic.

One difficulty with our subject is that man can study other things much better than he can study himself or his relationships. When we study ourselves we find difficulty in remaining objective. We domesticate animals, we cultivate plants, we store away goods for time of stress—but only a few of us can think of conservation with the general public welfare in mind. Rich natural resources in the hands of soulless monopolies may in the end prove destructive to the common welfare.

The pleasure of possession is so strong that not only individuals and groups of people but even nations are constantly clashing. Private interests and public interests are constantly at war. The haves fight the have-nots to preserve their gains. There is a constant struggle between private greed and public welfare. It is the age-old fight between good and evil, between the golden rule and the rule of gold, between construction and destruction as personified in Jehovah and Satan. Conservation is part of this struggle; it fights for those things which will benefit the greatest number of the present *and coming* generations. It is unselfish; its main aim is the preservation and maintenance of the general public welfare.

Conservation and Depletion

Our nation has been passing through a series of stages of depletion. We began with an abundance of many things, both living and dead. The supply of things such as wood, food, and minerals exceeded the demand. There was an over-supply of almost all natural products, and in consequence they were of little value. Men were profligate in their use of the gifts of nature.

Modern industries, chastened by the mistakes of our ancestors, strive to utilize waste and by-products although there must be some waste. Natural wastage due to wear-and-tear cannot be avoided altogether; it is only reckless, shameful waste that conservation tries to eliminate. In time, of course, our minerals will be exhausted; meanwhile all we can do is to use them without unnecessary waste and when they are gone find substitutes.

Another thing about waste is that it is a relative term. I have seen empty bottles and old newspapers traded for coconuts. I have seen men fight for a pinch of salt—which now sells for a few cents a pound. During World War II, African natives salvaged metal parts from wrecked airplanes and fashioned knives that they sold to American soldiers. Guerilla fighters in the Philippine Islands used old bottles, tin cans, and pieces of wire in building radios that contributed valuably to their struggle against the Japanese. Waste is a flexible term; it varies with time and place.

Although we may one day come to the end of certain mineral resources, with care we can preserve our living things. Conservation of this kind is tremendously important because no one has ever found a substitute for living animals and plants. Experiments may produce substitutes for the meat and eggs of birds, but where is the scientist who can create a substitute for the birds themselves? Obviously, the time to preserve living things is while they are still sufficiently abundant to hold their own and go on propagating. If allowed to live, they will replenish the earth; but once they are reduced in number, living organisms lose their niche in the economy of nature and gradually, in spite of what we do, disappear. At this moment, as you read these words, some things are well on their way to extinction and other things are just beginning. This world consists of the leftovers of things that were, and the beginnings of things to be.

Just half a century ago the forest of this country was virgin. It was not only a storehouse of basic materials with countless uses; it was a great factory which continuously transformed the carbon and nitrogen of the air into solids useful to men. But lumbering and blundering left great masses of slash. Only the best part of the best trees was used, the bulk being left to rot or burn. Fire followed fire until much of the forest land became grassland. Some of the grassland was used to produce scanty crops, and some was regularly burned over to improve the pasturage. Eventually overgrazing followed. In many places erosion by water and wind denuded the land, and leaching further impoverished it. Men, searching for patches of rich soil, lowered the water table by overdrainage so that in time it became marginal-agricultural land. Finally it became wasteland.

Thus the face of the earth was scarred. What is gone cannot be recalled, even though some restoration, upon which our future depends, can be accomplished. We must work with nature, not against her. We must not fight the force that gives us life.

What Can We Do About Depletion?

Much that has been saved was saved not by what we did, but in spite of it. Planning is vital to our future. Thoughtless lumbering must give way to forest management, which means working for forest betterment and perpetuity-of-yield, and constructing permanent homes for workers instead of flimsy ghost towns. It means providing woodlots on farms and establishing the proper kind of subsistence homesteads. It means reserving shaded pastures that offer protection and fodder for cattle. It means protecting wildlife. It means regenerating our wastelands for the general public welfare, saving together for the good of all. It means utilizing what some people call junk or waste, and other, more imaginative people find useful. These are some of the meanings of conservation.

We should preserve the beauty of the landscape and the health of the people. We should strive for a new kind of functionalism—to produce the simplest, cheapest, and best. Men should work constructively. Those who gather the "useless" things around them and transform them into something of use and value add to their own wealth and to the wealth of the nation. Man is merely a custodian of many things and when he passes on he leaves everything behind, even the ruins of what he destroyed. What a man does, both evil and good, lives on after he is forgotten.

A Philosophy of Use

Human resources, it seems to me, are the greatest of all resources because man to an awesome degree controls earth's other creatures. Man, however, is not supreme. Father Time and Mother Nature rule man, thus rule the world. They obliterate us whenever and wherever they will, with small concern for *our* plans. We are all subject to biologic law.

Nature decrees that man must work despite his innate love of leisure. Organs that are not used degenerate; hence use is the dictator of our being. Hand and brain alike must be used. When they are not, disaster may follow. People that have too much grow smug, become weaklings, and are superseded by the strong, who are themselves then caught in the endless cycle of decadence. Many of us can outlast adversity, but only a few can survive prosperity. Perhaps man is merely another one of the millions of experiments that nature has been trying on the earth. If he is fit, he will survive; if not, he will pass into oblivion just as numberless other creatures have. Through intelligent understanding of nature man can keep himself out of the class of "living fossils."

Men must maintain the fertility of the soil, the beauty of the landscape, and healthful surroundings. They must pursue constructive work in the production and processing of useful goods. Above all, men must work. We are all ruled by natural laws—the laws of supply and demand, of compensation, of diminishing returns, or of environmental fitness—none of which has ever been repealed. The man who breaks a law, either natural or man-made, should not thereafter expect its protection. The outlaw travels a rocky road.

I would say that one of the laws of nature is that he who scorns her bounties will in time suffer her furies, that making the most of our resources is today's number one problem. Without conservation there can never be any real progress.

From *Living by the Land,* pp. 7-8, 10-13, 15

2 The Florida Keys and Great Barrier Reef

If you try the Tropics at all try it by the sea where the air and water are pure and stimulating and where the products of both land and water are the builders of real bone and blood and imaginative and adventurous spirits. Rehabilitation of the Floridan Keys, pp. 66-67

The French for "the Keys" is Aux Cayes, the name of a city in Haiti, well known in early days for the quality of its rum. The word key, or cay or caye means little island. We have two types of keys, the outer and inland keys. The former border on the sea and may be divided into upper and lower. The inland keys and the lower keys consist of Miami oolitic limestone. The upper keys consist of coral limestone.

Key West was once the largest city in Florida. The shore keys were once thickly inhabited producing pineapples, limes, melons, grapes, vegetables of fine quality, without fertilizer or sprays, and products of the sea in great abundance. This is a loss to Florida and the nation at large. Even the date can be grown there by the side of the cocopalm, two of the world's most useful trees. Millions of good American money have been spent on three little islands in the West Indies with little effect, the total area of which is less than the size of Key Largo alone. Fully two-thirds of these West Indian islands are now unproductive. In far-off islands away from the watchful eye of public opinion and public press the dangers of misrule magnify with the distance. If they have good governors the chances for prosperity are good since virtue, like water, flows downward. Florida fortunately has a good governor.

Key West is certainly one of the important steppingstones to Pan-America. Like many other similar West Indian cities it successfully fought pirates, epidemics, hurricanes, and other difficulties but could not withstand the effects of a busted boom and the abandonment of an old well-established naval station. The rehabilitation of Key West is now in process. It is a quaint place with the best climate in the United States of North America. If properly developed and with the surrounding waters and near-by Keys scientifically, economically and artistically handled [it] will come some day into its own. It has a back-country in the surrounding Keys and shallow waters if properly conserved and not constantly rifled by reckless sea-rovers.

The mainland keys and the keys by the sea should all be treated as one, although the upper shore keys are of coral, the lower keys are the same as the mainland. Big Pine Key is covered with pines and has the same general appearance. The hammock or hardwood growth is very much the same on all. In fact this southern end of Florida including the keys by the sea is typical of the whole Antillean Area. Years ago when I first came to Florida the Perrine Grant Homestead Region and Pine Key in the glades were to me a sort of *ultima thule,* Latin for *"the jumping-off place."* It was a great level rocky plain covered with pines and palmettoes with here and there a swale of red soil or narrow glades connecting with the big Everglades beyond. We passed Seminoles with venison on their backs bound to market to barter meat for liquor. We reached it by wagon, usually an open work-wagon without springs, over a rocky road winding through the pines, a weary, lonesome and jolting trail through a mysterious but fascinating country where homesteaders, lured by the climate and free land battled with the forests and rocks. Here and there were dense hammocks of tropical trees. Now and then we stopped to rub our horses with bad smelling tar to keep off the swarms of horse flies, or to kill a diamondback rattler that might be sunning himself in the middle of the road. Here and there were the log cabins or shacks of the pioneers. There was venison, also gophers (a big land turtle) and comptie for food. Old settlers often argued that with the proper kind of fertilizer they could raise vegetables and fruits as good as those on the shore keys. These people envied the Key Conchs who had plenty of fish, vegetables and fruits, and at times, like manna from Heaven, they could have many luxuries from the ends of the earth when a ship was wrecked upon the shore.

People are at first staggered by the rockiness of this region. It is probably a blessing in disguise. Anyway the longer you live with it the better you like it and few old settlers would ever trade it for various reasons for white sand or black mud.

There is always a controversy as to what constitutes soil. Some say only what an ordinary plow turns over. Under this is sub-soil. The forester says the soil is the part of the earth penetrated by tree roots. Call it what you will, rock soil, or sub-soil, it is all the same to the forester if it produces a quick growth of trees and permits the roots to penetrate where they can secure a plenitude of water and mineral fertility. In choosing a soil the old forester looks up and not down. When you see, as is common throughout this country, tall pine boles, only eight or nine inches in diameter and sixty or more feet in height, or rich hammocks including such woods as mahogany, you can rest assured that there is something pushing as well as pulling them skyward. The main purpose of the soil is to hold the tree in place and furnish it with water and that is just what this rocky soil does. When you tell a new comer that he can get rock for house walls, rock for fences, rock for roads, rock for lime, also wood, sand, comptie-starch and water off the same wild acre he is incredulous and well he may be.

Dinner Key

When I first knew Dinner Key thirty years ago it was a little island projecting into Biscayne Bay. It was only a few inches above the water-level covered with mangrove above which towered three or four lofty cocopalms. It served as a landmark for boats passing up and down the Bay. It was separated from the mainland by a marl prairie a few acres in extent where the natives grew winter vegetables when the land crabs permitted and when free from water and mud. On this prairie the various rival baseball teams played on holidays and Sundays. The rough old rock-road from Coconut Grove followed the edge of the highland on to Miami, a narrow white way through a tunnel of green in Brickell's Hammock. This road on the land side was bordered by rough grey masses of Miami limestone which gave to the region the name of Silver Bluff. In the early days people picnicked on this little island. There were always plenty of dry buttonwood and driftwood in the neighborhood and boats on their way from the lower homestead region found this island a convenient stopping place for their midday meal. In that way no doubt it was gradually and finally named.

The next epoch in its history came with the World War. The shallow bay was dredged and the prairie was filled. Great masses of concrete covered the ground. Huge hangars were built and for a time it was a busy place training aviators for over seas. The old concrete water tower still stands. In time its work was done. The great structures of iron and wood were demolished and it lapsed again into quiet for several years preceeding its present and probably final service to mankind. It is now rated by some as the World's greatest airport. Thus gradually over a period of years Dinner Key has become a sort of hub from which radiate spokes of quick communication to many foreign lands which only a few years ago were very far away.

With the changes in Dinner Key the World has also changed. When used for a picnic ground it was the age of the slow moving boat and buggy. Today Dinner Key is in marvellously quick communication with islands to the south of us by the speeders of the air. These changes in communications call for changes in other ways so that speed and close communication may have beneficial influences other than exchange and trade. Dinner Key in all its changes from a quiet, wild, peaceful island on the shore of a quiet bay to its present full turbulent development is an epitome of what is happening to the world at large.

[In 1934 Dinner Key was the northern terminus for all South American air routes. No visit to Miami was complete without going to Dinner Key to watch flying boats like the China Clipper and the Yankee Clipper take off and land. The giant clipper ships played a major role in making Florida the gateway to South America and the Caribbean. In 1937, 40,000 people flew from Miami to the Southern hemisphere. Since then, the Miami airport has been relocated, and Dinner Key, still paved with concrete, is a large marina. The hangers have been turned into a convention hall, and the City of Miami has its administrative buildings there. Ed.]

The Great Floridan Barrier Reef

Through the kindness of a friend I sail by plane to Dry Tortugas. We follow the tail of Florida to the tip end of its last hair in the Gulf of Mexico, sixty-six miles west of Key West and three hundred and forty miles east of Yucatan. It is all a weird country, beginning with a real high land, houses and trees at Key West, passing over what seemed neither land nor water, tapering off into nothing but a storm tossed sea. It is truly nature's own aquarium finishing in the little archipelago called Dry Tortugas, neglected, unprotected and forlorn. For a panoramic view of nature's great handiwork there is no other way except the plane. Here are the bones of old wrecks, loggerhead sponges, tideriffs, schools of many colored fishes in waters of many colors, miles of white mud, atolls, sand-beaches and birds galore.

Although surrounded by water it is called Dry Tortugas. There are three things of special interest; the great fort, a mass of many thousands of handmade, well-laid bricks, hexagonal in shape surrounding a court yard with a moat on the outside. On the inside brick buildings, flimsy and wrecked, built hastily during the Spanish American War in sharp contrast to the solid, lasting work of long ago. There are beautiful sea-gardens near, and islands where thousands of noddy and sooty terns come hundreds of miles to nest each year. They have developed a marvellous homing ability.

We should clear out the old buildings in the courtyard and throw them into the sea. We should beautify this courtyard and renovate the old fort. We should sell the old iron to junk dealers, because this is a great historical curiosity worthy of preservation.

The old fort was built in 1846 with slave labor, now a dark, ramshackly wreck sizzling in the tropic sun like the Citadel of Christophe. This should be a part of the Everglades National Park. It should be renovated and shown as one of the great historical relics of the old south.

On the way there is a little group of islands called the Marquesas Keys—just why I have never heard. They are only twenty miles west of Key West. These little low uninhabited islands are surrounded by quicksands, and the whole region is and has been a graveyard for ships, bitterly exposed to all the changes of current and wave. In time of hurricane the whole region is churned into a mass of foam.

About fifty miles west from Key West is the Rebecca Shoal Light. And farther on is the Loggerhead Light on Dry Tortugas.

Few of us realize that Florida extends so far west of Key West in the form of a great barrier reef with little islands just hatching from the mud along the way.

In the study of bird migration, in the study of animals and plants of the sea, there is probably no more famous spot in the United States of North America.

A ship may get hopelessly buried in the sand, a small island may form around her, mangroves and other sea plants, together with other flotsam and jetsam may start, and then the whole thing may be washed away again in another storm. The great forces of nature have been battling on these islands and shoals, building and destroying, through the ages. It is our greatest barrier reef.

If the whole region could be completely blown up it would save the Government money in the maintenance of light houses, and re-move a menace to navigation, but it would probably change the vol-ume and speed of the Gulf Stream, which in turn might change the climates of countries bordering on the North Atlantic. It is a land in the making and little mangroves clinging to a mud bar may be seen out of sight of land.

In the south seas there are many picturesque islands called "atolls." They are looked upon by travelers as one of nature's masterpieces and have been studied by many scientists in times past. These rings of islands, with sandy beaches, with quiet waters inside, with openings here and there, with the homes of the natives under the palm trees, living in quiet bliss unmolested by worries, and unfettered by laws, have been for ages the dreamland of many restless souls. Like a rum-bling monster mill the waves grind the shells and pieces of coral into coral sand. This is blown inland in the form of spindrift and then later hardens into a soft limestone rock to be disintegrated again by the acids of vegetation.

A careful study of the chart will show that these Marquesas Keys form a perfect circle enclosing a shallow lagoon, the only natural atoll that I know of north of the Equator in the Atlantic.

The Dry Tortugas, from the appearance of the shoals, are or have been forming also into a circular shape. This archipelago consists of seven low islands of shell and coral sand. These were named by none other than Ponce de Leon, who in June 1513 captured one hundred and seventy turtles on these beaches. It was long famous as a place for the capture of turtles and turtle eggs, as well as the eggs of various sea birds that nested there by the thousand. It should be preserved as a breeding place for these creatures of the sea if for no other purpose. It is now inhabited only by lighthouse keepers with their families. This was at one time no doubt a resting place for pirates and buccaneers, and is still the place where suspicious looking Cuban and Mexican rovers come for reasons of their own.

Years ago I pried the tile from the floor of the engine room of an old English wreck. The coral polyp had formed a hard crust over half-an-inch in thickness on these tiles in less than a year. The master builders in these regions are the coral polyp, the mangroves, seaweeds and other marine life.

In the course of time an island forms out of the flotsam and jetsam of the sea. It is soon fit for the seagrape and the cocopalm. Other seaside plants are fighting the elements of nature and paving the way for more and better things.

There are but few places in the world more uniquely located. The great stream of water coming through the Strait of Yucatan and then on eastward into the Strait of Florida washes its shores. The prevailing east wind blows floating objects on its beaches.

The Marquesas Keys and the Dry Tortugas are marked on our State geological map as being of Miami oolite, the same as the islands around Key West and the mainland.

The keys north of Bahia Honda are classed as Key Largo limestone, the one is a white oolitic limestone, the other coral reef limestone, in great favor for high class construction.

At one time Key West was the liveliest city in Florida. Its big business for years was wrecking and the region of the Dry Tortugas was a fertile field. Although the laws were often violated, wrecking was a legitimate business. A considerable percentage of the salvaged material went to the wreckers, lawyers, auctioneers, wharf owners, etc. It is often told how the wreckers were in league with the ships' captains, and how many ships were purposely wrecked, but the worst story I have heard was the case of the ship "America" wrecked on Dry Tortugas in 1859, when the lighthouse keeper, in league with the wreckers dimmed the lights to fool the captain. Licenses were issued to reliable wreckers, but the courts were busy settling disputes which naturally arose out of such an adventurous business. This was truly a graveyard for ships and the wreckers and courts were the obliging undertakers.

Since the Marquesas and Dry Tortugas belong to the Federal Government the same should be incorporated in the Everglades National Park. This would preserve it for all time so that it could be used by students of the future.

Of the fifty or more hurricanes of which we have record, fully half passed into the Gulf just west of Dry Tortugas. It is therefore an important weather station.

Southward across the Gulf is also a favorite bird migration route. The birds have learned to shun both Florida and Mexico just as they shun Italy on their winter visits to the tropics.

The coral keys from Bahia Honda northward have been elevated from the sea. They are of coral, and coral never forms out of water.

Canova* describes how many years ago there were curious green streaks of water in the Gulf, and how fish and sponges, in fact everything in its wake were killed. Cuban fishermen, in passing from Key West to Havana, had live fish in wells in their boats. These died when they passed through these mysterious green streaks. It was generally supposed that some poisonous gas had come from a crack in the bottom of the sea. Only recently dead whales were washed on the beach at Cape Sable, whether sick or scared or what, their decomposing hulks filled the air with bad smells and their bones are still whitening in the tropic sun.

To the south of these keys and shoals and to the west the water is very deep. To the north along the coast of Florida, it is comparatively shallow fully a hundred miles into the Gulf of Mexico so that the Dry Tortugas, the Marquesas and other keys and shoals form a reef-like dam on the south edge of this shallow area through which there are many channels where the water rushes in and out with great rapidity and volume.

It is, in short, a great barrier reef of much interest to the scientist and a source of never-ending joy to the lover of the sea and of the gambling sweepstakes which it drops at your feet.

[*Canova, Andrew P., *Life and Adventures in South Florida*. Tampa, 1904. Ed.]

Sea-Gardens

The word "garden" has few peers in the English language. Whether in the word itself or in what it represents I am unable to say but of all our words none is more euphonious and few more frequently used in song and story. Whether it is just a kitchen-garden or market-garden or botanic-garden or hanging-garden or sea-garden it has the same respectful significance, although probably of common origin with our word "yard." It usually implies an enclosure, although is often applied to broad unrestricted stretches of wild flowers and to sea weeds, corals, bright colored fishes and other beautiful things on the pastures of the sea in the transparent parti-colored waters of tropical shores. In no place are these sea-gardens more beautiful than on the Florida Keys and in other similar places throughout the West Indies and the Spanish Main. These sea-gardens are never protected. They belong to everybody and everybody helps themselves. When the Everglades National Park becomes a reality it will include many miles of sea-gardens which should be protected. It will be the first and only tropical marine park in this country.* If, as I always contend, there is no greater asset in a tourist country than natural beauty these sea-gardens are worthy of special attention. To properly describe these gardens of Neptune in these pellucid waters would exhaust my supply of superlative adjectives.

Seaweed is a great source of fertility swept in great quantities on the shore. The Japanese convert one of their seaweeds into the famous agar-agar used throughout the world for many purposes. Californians use their kelp in the manufacture of iodine.

There are many kinds of sponges in this region and for many years it has been an important industry. There are few better sponges than the Matecumbe sheep's wool. Mata in Spanish means "bushes," combar means "to bend." Maybe this interesting old word Matacomba means "bent bushes." Surely a picturesque and appropriate name for an island where the bushes were no doubt bent by the unobstructed sweep of the wind from the Gulf across to the Strait of Florida or vice versa. On Largo there was once a beauty spot called Garden Cove.

Down on these Keys amid the brilliance of coloring and wonderful clarity of briny waters there is a peculiar picturesqueness about this old Florida sponge industry in spite of the stink which hovers over it.

[*John Pennekamp Coral Reef State Park was finally established in 1957. Ed.]

Everywhere throughout this region you will see the sponger with his sponge hook and water glass looking into the still water for the ugly grey masses on the bottom. Of late the Greeks have come from the Ionian Sea with brightly painted boats and diving suits to compete with the old timer on his home grounds where he was once also a wrecker and a grower of sours and dillies on the rocky Keys. They were always watching for wrecks which a kind Providence might send in the way of succor at any time. Then the Conch horns blew and as if by magic the ship would be surrounded by many boats and helping hands, helping themselves to everything that they could safely move.

This country is biologic headquarters where marine life can be studied in many forms and where the sponges, the fishes, the shellfish, turtles and other things can be properly fostered and protected. On these little islands swept by wind and wave the mahogany, the prince of hardwoods, the yardstick of quality for cabinet woods, and the lignum vitae, the world's toughest wood, and ironwood, the world's heaviest wood, will survive and flourish if left alone.

Above the smell of the curing sponge the pungent smoke of the buttonwood floats over the water while the sponger broils a crawfish or cooks a stew of conchs. In case of a gale he seeks a sheltered cove amid the storm-safe mangroves. When the catch is done and the sponges are beaten clean and washed many times he sets sail for home, his beloved Key West, the capital of his work and dreams.

There are many things that render the Florida of old precious to those who have long lived here. The Seminole with his little family in his cypress dugout on creeks in the glades, the clouds of birds in swamps of cypress or mangrove, quiet lagoons fringed with palmettoes and cocoplums and last but not least, green keys in particolored waters with the Conch at work in his picturesque boat by his sponge or turtle crawl. These and many other pictures the old-timer will carry in his mind's eye as long as he lives.

Perhaps these beautiful sea-gardens on our reefs consisting of countless living things busily building their homes of lime rock are having far-reaching effects. When the great storms come and open up new channels in the masses of mud and marl like a mighty broom the ocean floor is swept clean again. Out of the sea all things once came and back to the sea they will in time all go. Even the bones of men that go down in ships are in time completely absorbed. Coral-rock heads grab passing victims and when a ship is buttonholed she is held for keeps, while in the sand she is slowly smothered like the octopus absorbs its victim. In the slime of the sea the living creatures of the world had their beginnings. In fact in no place on earth can nature in the raw be better studied. When we look into the rock holes on these coral reefs we see primal and peaceful gardens of beauty.

In short these sea-gardens, clam beds, coral reefs, sea islands, with all the other forms of sea life, constitute next to climate, our greatest resource. It is a lure that captures the imagination of all of us. It puts us into a primitive state of repose where we belong. It links us with the genesis of the world of which we are really only a very small part and by no means the master. The combination of sea island, sea-garden and sea atmosphere furnishes food for soul and body. Many of these little islands are merely door steps to the gardens of the sea. An eminent scientist once said that man should be raising thousands of sea cows on sea shore pastures. We need today a study of all the resources of the seashore, the propagation of oysters, clams, conchs, fishes, sponges and a host of other things that add food and comfort to mankind. Few of us appreciate what a tropical sea brings, only those of us who have lived by it for years know its smiles and frowns. In northern climates trees shun the sea, but in the tropics I have seen valuable food products in the form of fruits and vegetables grow so close as to be drenched with salt spray. This area like the rest of the Antillean Littoral will some day be precious because of its limited area, accessibility and potentialities. No man can starve in such a place. Foods from the sea and near the sea have quality. They contain minerals that feed the brain, bones and glands. The briny water is like balmy oil to the skin. Artificial fertilizers are unnecessary and in times of stress nature beckons us back to where land and sea meet to the spot that is primal and constantly regenerative. Those who stick to inland pastures and weevil-infested cotton fields are wasting their best opportunities and road builders that refuse us access wherever possible to these little islands of the sea are depriving us of the main source of life and vigor.

When I refer to sea-gardens I mean not only those natural beauty patches in the reefs by the Gulf Stream but in real gardens on the sea bottom for the cultivation of sea truck and the development of these little islands to furnish mankind a terrestial footing close to the scene of his operations. Soil quality is secondary. Location is of paramount importance. There is no soil, no matter how rocky, in the tropics which cannot be rendered productive, especially by the sea where there is an abundance of natural fertility. The sea not only furnishes it but at times throws it at you. The most valuable part of the southeastern United States is Florida. The most valuable part of Florida is in its southwestern archipelago. No time should be lost in rendering every inch of it available to everybody under wise restrictions as far as possible. Roads to muck and sand only render muck and sand available. Roads to the sea render the sea and all that it brings, an open Sesame to everybody. For the future look to that neglected territory to the south and west of us. Key West was at one time the largest city in Florida. Indian Key was one of the oldest ports and trading posts in the United States of America. The Keys at one time supported a large population before the mainland was settled. History repeats itself. They will come into their own some day. With a little soil sprinkled over the rocks, with a slat house for shade and wind protection, a garden patch on the Keys will produce a larger quantity of the best quality of fruit and vegetables possible in the State of Florida. Right by your door also you may have a sea-garden, the like of which cannot be found elsewhere in this country. If you try the Tropics at all try it by the sea where the air and water are pure and stimulating and where the products of both land and water are the builders of real bone and blood and imaginative and adventurous spirits. And if the government is anxious to help there is opportunity for constructive work as well here as in the Tennessee Valley or in the deserts of the west. But the work should begin at home on the Peninsula of Florida and then gradually extend onward southward, eastward, and westward with the proper connections and relationships.

These Keys have been for many years the playground of presidents.

From *Rehabilitation of the Floridan Keys*, pp. 15, 18-21, 26-31, 45-46, 62-67

3 The Everglades National Park

. . . Where the crocodiles, roseate spoonbills, ivory-billed woodpeckers and other creatures of land and sea may rear their young without molestation.   Rehabilitation of the Floridan Keys, p. 37

Over the mainland in the region of Cape Sable is the territory destined in time to be a National Park. This area will in due course come under the control of the National Park Service.

Over on the mainland of the Gulf due west of Miami is this wild country little-known except to adventurers, sportsmen and renegades. In one place are the ruins of an old sugar plantation, in another the rusting machinery of an old tanbark mill, and still in another even the site of a boomtime town. The forces of nature were too violent for human contentment. The sheer wildness of the country excludes mankind except for a transient sojourn. After a visit of a few days you emerge from its fastnesses bewildered by electric lights and the speed of autos. Like many lost lands of the tropics it sleeps in peace unknown and alone except for the roar of wind and wave. Near and yet so far, however without doubt it is one of the strangest and wildest of tropical shores, with countless islands and countless watercourses affording drainage to that sea of saw grass in that bowl of muck called the Everglades.

Your loneliness is not entirely due to the multitude and complexity of islands and creeks. Much of it is neither land nor water. Sometimes it is one, sometimes the other. It is lonely because it is strange and wild. Many are beset with fear—fear of being lost, fear of the "madness of the wild." On the west side it is bounded by the Gulf of Mexico, which is not always quiet and harmless. Besides there are many beds of raccoon oysters close to shore which are jagged and bare at low tide but dangerous to navigation at all times. It is considerable distance from settlements in all directions. It is secluded by miles of muck, marl or jagged rock with dangerous pot-holes. A highway from the east to navigable water on any one of these rivers would give access to this whole region, since the majority of these streams are connected by many cross channels not shown on any map. I refer to the thousands of islands more or less, the so-called Ten Thousand Islands south of Pavilion Key, backed by a great delta-like territory fit mainly for the home of the countless wild creatures, plant and animal, that are native there. I have ascended all its rivers and can think of no other uses to which it could be profitably put. Under the head of land-use economics here is one piece of land about which there could be no controversy. It is fit simply to marvel over, and fit for the home of creatures that once lived there in countless numbers. It has in addition a great scientific interest. It is natural, unaffected by man's interference, one of the few places on earth that we can hand to posterity without mutilation if we hurry in the process of preservation. It is especially fit for a tropical aquatic park to be always reserved just as it is for recreational, educational and preservational purposes.

This archipelago consists of patches of mangrove in a labyrinth of streams through which the tides sweep in and out by many beds of raccoon oysters. The mangrove crawls like a crab on its many legs. The stilt-like branches reach down into the briny water and grasp the bottom. The tree is firmly fixed in the muddy flat and like a net of wire catches the flotsam and jetsam which come its way. Whether a coral strand or scraggy oyster bed, if it once gains a footing it marches forward and holds its own against wave and tide. It has its willing help-mates in the form of the black mangrove, the white mangrove and the buttonwood. The black mangrove sends up countless stick-like roots to form a solid mat or brush-like surface to the soil. The red mangrove is the forerunner. It is a plant with a purpose. It is the

consolidator and benefactor of many acres of mud, with oysters clinging to its roots, where even the fishes feed among the fallen leaves,
where even the shallow draft boats may navigate. Yet in its branches
orchids bloom and birds nest. Ten Thousand Islands—there may be
more, there may be less—they have never been counted. It is only a
guess. Anyway, they are multitudinous and very much alike.

As you ascend one of the many rivers you soon leave the salt and
brackish water. You leave the sea, and also the mangrove and many
insects. The vegetation changes. The water is clear, cool and drinkable.
Alligators, snakes and turtles glide into the water. The sound of the
engine echoes throughout the woods. Water fowl fly ahead up the
river. The shores are lined with hammock trees. Some of them, such as
the magnolia and maple are native to more northern regions. One
beautiful vista succeeds another. The river grows narrow and it looks
like the end, but suddenly, unexpectedly widens again into a broad
beautiful bayou. There are perfect reflections in the water. Fish are
jumping here and there. Now and then a cabbage palmetto is dead
where some bear has eaten the tender tip. A little deer may be seen
browsing in freshly burned areas. We stop at a Seminole home. It is
empty. Bananas and guavas, unpicked, are growing in front of it.
There are no bundles of any kind left tied under the rafters. No doubt
there was death in this house and they left it to burn or blow away.
He or she passed on into Hopie Land, and somewhere nearby there is
a grave containing the body and belongings of the dead. There are
many shell heaps and kitchen middens of a departed people.

Soon the hammock on the bank fades into sawgrass. We have
reached the Glades. Here and there are islands of silvery-grey, big-
buttressed cypress. They are short in stature but very grey and old.
Their tops are filled with airplants. It is winter and they are bare of
leaves; soon they will be covered with a mass of pale green. Here and
there are patches of the cabbage palmetto. Here too is the dark leaved
cocoplum. There are clouds of birds of many kinds. The white egrets
look like a mass of flowers in the dark green foliage.

At times the Ten Thousand Islands are cursed with swarms of flies
and mosquitoes. It is one of the world's greatest fly and mosquito
centers. This decreases as you ascend the rivers. There are few insects
in the Everglades proper. There is no stranger scene than the nuptial
flight of the horse-fly at dawn.

Then we turn back to the salt again and up another river. The same pictures repeat, the same scenery as before. It is all a wild reservation where man has no place except as an onlooker. Much of it is in the making, much of it looks unfinished. It is a series of peaceful pictures combining to form a type of scenery not common elsewhere.

The picturesque Seminole with his little family, all brightly dressed, poling his dug-out belongs here.

As to its history much could be written if we knew the truth. Old Spaniards knew of it long ago. In fact this region was once called Ponce de Leon or Chatham Bay. It was traversed by the hunters of Indians who were paid a bounty for man, woman or child.

In times of storm it is a place of refuge. The trees along the streams break the force of the wind. The quiet water is fresh and pure. Meat and fish are plentiful. The Ten Thousand Islands, with their fish and clams, are near. It would be the most perfect place in all the world to seek peace and quiet and plenty of food but for the bugs. Instead of a haven of rest away from hustle and bustle it becomes at times a place of torment. There are tropical beaches of the Cape Sable region where in the great mill of pounding waves the rattling shells are ground into sand. There are hammocks where the mahogany, the prince of hard-woods grows, also groups of patrician royal palms and many other rare tropical trees. There is the great barrier reef of sand, muck and coral which extends seventy-five or more miles west of Key West into the Gulf. There are the Marquesas Keys and Dry Tortugas, which resemble the idyllic atolls of the south Pacific, and the Great Bank of clams on the West Coast, the largest bed of the largest clams in this country, containing two hundred square miles of chowder and fish bait. There is also the great Wall of Mangrove bordering on the Gulf, a queer association of trees that live in salt water and fight the sea and many rivers and their connecting branches leading inland till they are lost in the saw grass. All of this will soon be a National Park, a sort of show window for South Florida, where the crowds that come as time goes on will see something that they have never seen before and something that they cannot see elsewhere farther north. It is not alone for scientists and pleasure seekers, but for all who are eager to see and learn.

This is the only way to put this vast area to a wise and profitable use. No one can speculate with it. It will be off the market forever.

The purpose of this park will be educational, recreational and preser-vational, for all the people and not for the plume hunter, the pot hunter, smuggler or renegade. The wild life will be protected and many rare birds and other animals will be saved from extinction, and some of the rare things which once lived there, such as the flamingo may return. This applies also to rare tropical trees and other plants not found elsewhere in the United States of North America. As time goes on the eyes of the world will be turned toward it just as they look at the great reserve with its many wild creatures on the banks of the Limpopo, or the Yellowstone. It will be a place to go when the rest of this country is snowbound. It will be one place where the crocodiles, roseate spoonbills, ivory-billed woodpeckers and other creatures of land and sea may rear their young without molestation. And there too the mangrove islands, mud flats, beds of raccoon oys-ters and clams, sponges, conchs and corals and all the other things that inhabit the sea and the shores of the sea may form and reform with-out destructive interference by the hand of man.

From *Rehabilitation of the Floridan Keys,* pp. 32-37

4 Dr. Henry Perrine

. . . one of the real founders of South Florida.
                    Rehabilitation of the Floridan Keys, p. 14

Much has been written about Dr. Henry Perrine. The sensational details of his massacre have been stressed to such extent that his real accomplishments have been obscured. Suffice it to say that he was killed by a bunch of drunken Calusa Indians. They were Spanish Indians and probably had little to do with the Seminole War or the Seminoles. They were real pirates out for liquor and loot.

The idea of the small tropical subsistence homestead was uppermost in his mind. Some of his statements of over a century ago apply as fittingly today as they did then. In the formation and development of the Perrine Homesteads we are, after long delay, applying the same basic ideas that Dr. Perrine had in mind when he asked for the Perrine Land Grant for colonization purposes and for experimentation with tropical and sub-tropical crops.

Florida was not at that time a State. March 20, 1845, five years after Dr. Perrine's death, both Florida and Iowa by the same Act were admitted to the union. In the same year Dr. John Gorrie of Pensacola discovered a process of artificial ice manufacture. Although the Seminole War was never officially ended on any particular date there was trouble with the Indians until 1858. There was a court at Key West for the whole of South Florida as early as 1828. The following quotation is as applicable today as ever. It is attributed to Dr. Perrine. "The character of the vegetation in tropical Florida will ultimately create a very dense population of small cultivators, and of family manufacturers of numerous diversified products, which will thus prevent excessive overproduction, or ruinous rivalry, in any single branch of culture or of manufacture." You will note he recognizes South Florida as a tropical country and goes to Yucatan to get suitable plant material. He stresses small holdings, diversified crops and home industries. I have used the very same words in the above quotation so much that I might be accused of plagiarism were it not for the fact that these ideas have been rolling down the ages unheeded from time immemorial.

An Act of Congress was approved July 7, 1838 "to encourage the introduction and promote the cultivation of tropical plants in the United States" and a township of land was granted to Dr. Henry Perrine and his associates to be located in one body six miles square upon any portion of the public lands of Florida south of $26°$ north latitude. The act provided that whenever any section of land shall be occupied by a bonafide settler, actually engaged in the cultivation of tropical plants a patent shall issue to said Dr. Perrine in making proof thereof to the Commissioner of the General Land Office.

When Dr. Perrine was killed these rights and privileges were vested in his heirs. Thirty-six families from the Bahama Islands settled there but the majority moved away or scattered because of the turmoil of the times in that region. Several of the sections were settled and planted to tropical trees and the land was finally granted to the Perrine heirs and his associates. It is still known as the Perrine Grant and the little town of Perrine is a very active fruit and vegetable center and it is generally recognized that although Dr. Perrine had plenty to choose from in those days his choice of that section was a wise one. Those familiar with South Florida have no hesitation considering all things in asserting that there is probably no better section for small tropical homesites.

The main object of Dr. Perrine's endeavors was to introduce tropical plants into tropical Florida on small homesteads to furnish crops for home use and supply materials for the maintenance of small home-steads and for small home industries.

The native plants of South Florida were tropical, the kinds carried by winds and birds and the original natives were of such a wild nature that they did very little in the way of agriculture. They no doubt introduced cotton, also several species of cacti, the fruits of which they relished. The Spaniards probably planted the cocopalm although on the plat of Indian Key showing what Dr. Perrine planted there are several palms of some kind.

Sisal for rope manufacture was one of Dr. Perrine's strong points of endeavor and this idea no doubt led him to Yucatan for his materials. Among the Indians the fibre was used for ropes and hammocks and he of course at that time did not realize the great future use of the fibre for binder-twine in reapers and binders. The sisal we have here is no doubt a descendant of the plants he introduced a century ago. In order to establish plantations elsewhere slips were supplied from the Florida Keys and as late as thirty years ago I was on an old boat that carried sisal slips to Nassau.

He brought in mangoes, avocadoes, limes, sapodillas and also a mulberry for the growing of silk worms. He secured the appointment of consul to Yucatan for this very purpose. He sent his materials to a Dr. Howe who planted them on the Keys for a start with the idea of moving them later on to the Perrine Grant. His selection of Yucatan, although it may have been accidental because of his interest in sisal, was a matter of special interest to me.

The great peninsula of Yucatan is very much like Florida. It is however drier and in moving plants from a dry region to one with more moisture you are surer of success than vice versa. It is still one of the very best spots for us to look for useful plants for introduction into Southern Florida. It is a part of continental tropical America and has had a very old and high grade civilization. There are the same kind of rocks, the same kind of jungle, the same mangrove swamps and the same lime sinks there known by the euphonious name "cenotes." There are the same kind of mudholes and pan-shaped depressions called "aquadas" and "sartenejas."

The vegetation is very much the same as South Florida except that there are more species because it is part of the tropical mainland and because the Mayans with their great skill in many lines must have had some valuable cultivated plants other than sisal, chicle and corn.

Dr. Perrine was very active and had a real part in securing the money for the establishment of the Smithsonian Institution in Washington. He went to England with another person for this purpose and it was no doubt Dr. Perrine's sincere interest in science and his great enthusiasm that materially helped in this conquest since others had tried it before without success. This alone is a matter of great historic value not only to Florida but to the world at large. He has other things of equal importance to his credit.

It was in 1838 when the Florida Tropical Plant Company was organized. Dr. Henry Perrine was President and Charles Goodyear was Vice-president. It is claimed that Charles Goodyear was on Indian Key the night of August 7, 1840 when Dr. Perrine was killed and that he saved himself by hiding in a native rubber tree. Dr. Howe who lived on the Island and who cared for the plants shipped in by Dr. Perrine was spared.

One of this Company's plans was to grow rubber in South Florida. It was in 1839 when Goodyear dropped some rubber and sulphur on a hot stove. The result led to the vulcanization of rubber. These great men away back in those days had more faith in Florida's productive capacities than we have today. They realized the possibilities of South Florida along several lines and although these great businesses have been developed in regions where labor is cheap and the standard of living low there is still chance for it here. Those men were not all wet in their hunches. Rubber may be successfully grown in Florida some day. It would have happened long ago were it not for the encouragement our markets and capital afforded to slave labor, although synthetic rubber may soon again change the picture. Slavery was abandoned to accept in trade the goods still produced by a condition of labor equal to any kind of slavery anywhere.

The rubber tree they grow in the Far East is none other than the native Brazilian. The quinine they raise is the quinine tree of Ecuador, Colombia and Peru. They say there is still plenty of it growing wild in Colombia and Dr. Perrine was the man who introduced quinine powder into North America from France.

Few of us realize that the exploration not to mention the development of parts of the Tropics would have been impossible without quinine.

Only those who lived in the Tropics years ago appreciate the value of this drug. It is still necessary in many places and you are always safer with some of it with you. I remember once in Surinam how a Dutch trader sold a poor substitute for quinine. When the miners came back from the jungle full of fever they took the trader into the swamps, stripped him and tied him to a tree. I remember how my old professor always administered five grains of quinine no matter what ailed you. He was once a member of an expedition of five into the jungle. He took quinine every day. The others neglected it and he was the sole survivor. He always prescribed plenty of lime juice in the water you drink and a little quinine every day.

Getting the quinine tree seed out of South America was a difficult job. It was filled with more thrills and romance than story books. The old Peruvian Indians discovered it as well as many other things of great value. There are other fever barks but none so potent as good quinine. It might be grown right here in South Florida. I doubt if it has ever been tried or ever had a fair trial.

The full story of Indian Key where Dr. Perrine established his headquarters has yet to be written. In this day of easy communication we forget that in Dr. Perrine's time South Florida was practically an island. Even Fort Jefferson on Dry Tortugas was not begun until 1846. This part of the State was completely separated by miles of black mud and unbridged rivers. Within the memory of many now living the mail carriers walked the beaches and paddled across the rivers in small boats. The first members of our legislature from Key West went to Tallahassee by way of New York.

Our first port of entry was Indian Key close to Bahia Honda Channel. It was used by the English and the nearby water was called Spanish Harbor, all well fitted for the shallow sail boats of those days. Back of this island across the shallows of the Bay of Florida was the great hinterland unexplored except for wild Indians, runaway slaves and other renegades. The early associations of these Keys were with the Bahamas and Cuba rather than the rest of the United States of North America.

These Keys were all favorably located for wrecking, a profitable and exciting industry in early days. It is probably just as rich in story as the famous South Seas. When Chakika, the last Calusa chief, looted Indian Key he loaded his canoes and paddled into Shark River and on into the Glades to hide on Chakika Islands some distance west of Fort Dallas, the site of Miami which means in the Calusa tongue "Big Lake" the same as "Okeechobee" in the Creek language.

There is no spot in all Florida of more interesting historical background than Indian Key and its environs in that strategic archipelago known as the Lower Keys. If the Perrine Homesteads are successfully established there should be a monument to Dr. Perrine in commemoration of his work in general but in particular to the consummation of his original idea for the establishment of small farms with diversified tropical crops yielding food for subsistence and materials for the maintenance of many local industries. He was one of the real founders of South Florida.

It is a remarkable fact that so many distinguished men had something to do with the development of Florida. Many came to play. They even built orange groves for fun but Dr. Perrine was serious, in fact too serious, to attempt, in the turmoil of those days amid enraged and combative Indians as well as pirates and buccaneers, to start a settlement in the back country away from everything. The Keys were settled first because water was the only avenue of traffic, although fresh water was scarce, and because the pickings were good from the sponges, fish, and shell fish in the sea and the many wrecks on the reefs. Their favorite crops were limes and sapodillas, called sours and dillies by the early Conchs, both of which were brought no doubt from Yucatan by Dr. Perrine.

From *Rehabilitation of the Floridan Keys*, pp. 7-14

5 Beach Combing and Sea Salvage

Buying is never as much fun as finding.   Living by the Land, p. 117

My definition of beachcombing, and I think, the one commonly accepted, is searching the seashore for anything of value or interest that
has been washed in by the waves or blown ashore by the wind. Beachcombing is this and more, for the beachcomber carefully studies his
findings and identifies them if possible, performing with modern objects the same kind of work the archeologist does with ancient ones.
Beyond this, beachcombing can be a healthful hobby, inviting the
adventurous to long hours of tramping along the shore.

   At one time beachcombing was more than just a hobby or a pleasant form of exercise. Many years ago when those of us who lived near
the sea needed chairs for our homes, pots and pans for our kitchen, or
even clothes for ourselves and our families, we first scoured the beach,
hoping that we might find them there. If we came back empty-
handed, we reluctantly took down the mail order catalogs and sent
away for the articles. But buying was never as much fun as finding!
More often than you would think—especially after a wreck at sea—we
found scores of things for which we had use.

   Some people who lived along the shore had more impatience than conscience in those days and could not wait for providence to send them wrecks; they "encouraged" them. Even today tales are still told along the coast of unscrupulous men who lived from the spoils of wrecks they had caused. Not all the wreckers—the men who salvaged the contents of wrecked ships for a certain legal percentage—were dishonest; yet many a ship was lured onto shoals by false lights and other devices. In one place I know about, natives imitated the flashes of a lighthouse by cutting holes in the side of a barrel, placing a light inside, and rotating the barrel. At another place they tied bamboo poles to cows' horns, hanging a red lantern on one end of the pole and a green lantern on the other. Then on stormy nights they drove the cattle up and down the beach, confusing the pilots of the in-coming ships. Occasionally lighthouse keepers were bribed to falsify their lights. The wrecking pirates were so enterprising in some places, that they sent agents to big ports, where they bribed the captains of ships which passed their shores to stage an "accident" that would provide rich salvage. Now and then ship owners wrecked their craft in out-of-the-way places to get the insurance. In recent times, the increased use of radio and other modern warning devices, and the alertness of watchful insurance companies have reduced the pickings of honest and conniving salvagers alike. A colorful, but costly, occupation has disappeared.

During World War II, enemy submarine action off the coast of Florida resulted in a number of sinkings. Although much of the contents of these ships went to the bottom, a considerable amount washed ashore. Residents of the Florida Keys frequently found huge pancakes of rubber, wax, dried fruits, cooking oils, and other valuable materials on the sands, some of them in watertight containers. Even scrap iron and other metals were sometimes deposited by the sea on reefs and beaches.

No one knows what the sea might toss upon the shore. Masses of seaweed, sponges, seeds, bamboo, lumber, bottles, pumice and coral, shells, and building bricks worn smooth by the grinding of the great ocean mill—all can be found on the sand. I have seen dead men washed ashore, dead whales too, and once I saw a live horse carried in by the waves. Old automobile tires are often found. These have broken loose from the fishing boats on which they were used as fenders in place of the comparatively expensive woven rope fenders.

Severe tropical storms occasionally bring in enough seaweed to cover the land ten feet deep, completely smothering groves of limes, sapodillas, guavas, sugar-apples, and other fruit trees. The trees survive when heavy rains bear down on the seaweed, whose decomposition in time enriches the soil. Tropical storms of this degree of intensity can be expected about every ten years. Much of the storms' debris is torn from the bed of the ocean (sponges, for example), and some of it may come from the great Sargasso Sea, far to the east.

Seeds are brought in too, in foul weather and in fair. Hence, varieties of plants travel from place to place. *Avevectant* seeds are those brought in the stomachs of birds or carried in the mud on their feet; *aquavectant* seeds are those floated in by the water; and *ventevectant* seeds are those carried by the wind. Because the coconut and other aquavectant seeds enable plants to migrate over surprising distances, a large number of wild plants are common on both shores of the Atlantic Ocean.

Some of the grasses that grow along our coastline may have come to us long ago on slave ships. The slaves slept on coarse hay, and when the ships reached our shores the bedding was thrown overboard. It is probable that seeds of various plants were mixed in the hay and floated ashore to take root here.

I have often wondered at the number and variety of bottles that float in. Many, of course, sink before they reach shore and many are broken on the rocks, but, even so, almost every beach is strewn with bottles dropped from passing ships. Since the Gulf Stream, just a few miles off Miami, is one of the world's great sea lanes, ships are visible from shore almost every minute of the day and night. The number of bottles that can be picked up from the sand is a good indication of the amount of liquids drunk by people on passing ships. On one small beach I have collected empty bottles by the wheelbarrow load for use in building masonry walls, and I could easily speculate about the most popular brands of liquor used by ships passing the tip of Florida. Sometimes I have found bottles containing, in the best story-book tradition, messages. I confess some disappointment in my discovery that all the messages were intended as jokes or were deliberate fakes.

The great Gulf Stream is really a continuation of the Great Equatorial Current. The rate of flow is slow through the Straits of Yucatan, but it is accelerated off the Keys of Florida by the addition of much water from the great rivers that empty into the Gulf of Mexico. Close to the Florida shore and deep beneath the surface are counter-currents which bring fine sand from the North. Along with this fine sand, disintegrating sponges and other sea creatures containing silica are ground into minute particles. The sand is sometimes in granular form, sometimes in tiny spicules. Sometimes the granules of sand are surrounded by lime. Oölitic rock is usually part silica, and much of it is no doubt of eolian formation. Some of it is like a calcareous sandstone and some of it like silicious limestone. Some of it hardens quickly into rock and when burned forms good quick-lime; some of it turns to sand when burned. Coral rock and shells are almost pure limestone.

Quite a lot of driftwood washes onto the beaches. If not riddled by the teredo worm, it is as good as ever—in fact, better than the imperfectly seasoned wood available during wartime. One can usually collect enough driftwood from the shore to construct small buildings, and there is always enough for fuel. The driftwood burns with a multi-colored flame. Now and then a large mahogany log washes off the deck of a ship northbound past Florida, carrying its burden of fine wood to ports on the northeast coast of the United States.

Tropical woods that have been tossed about in the sea awhile usually become stained so that the beachcomber who tries to identify the wood meets a tantalizing problem. Once I found an unusual piece of wood on the shore. After examining it carefully, I was uncertain about its kind and decided to send it to an expert in New York City. He, too, was unable to identify it, but rather than admit his ignorance, he sent it to another expert in Washington, D.C. Unfortunately the Washington expert was unable to identify it and, without telling his New York friend, forwarded the piece to me. I replied that I thought it was blue-mahoe from Cuba, not letting the Washington man know that I was the one who had found the specimen in the first place. From Washington the authority on woods relayed my decision to New York, and shortly thereafter I received a letter from there informing me that the piece was probably blue-mahoe from Cuba. No one in this little farce admitted that he had asked the advice of someone else; it was like talking to myself.

Despite the amusement I got from the incident, I thought no less of the northern experts. After all, identification of the wood had been made difficult by the stains it had taken on in the water, and they had no authentically labelled samples for comparison, nor leaves, flowers, or fruit to make identification conclusive. Thus, in several well-known museums in this country I have seen mislabelled samples of wood. One of the professors of my student days in Germany used to finish his lectures with the statement, "Further investigations are necessary before conclusions are warranted."

On the mainland and the Keys of Florida the beaches are yielding riches to those who know how to find them. In the northern part of the state the sands are giving up ilmenite, a mineral consisting of iron and titanium; rutile, an ore of titanium; zircon; and monazite, a phosphate of cerium. The sands along the ocean may contain gold, but not in an amount or form to be easily recoverable. It is probable that buried treasure lies hidden in some of the sand dunes, and certainly many dwellers by the sea have been buried in the dunes, in coffins appropriately made of driftwood.

Along the shores many salt-ponds or shallow lagoons are banked. The sun evaporates the water, leaving behind salt which can be raked into piles and used in curing meat and fish. Iodine in the salt prevents goiter, a common ailment among inland people.

Many people who live along tropical shores had little interest in wartime rationing of foods. Red points and high prices mean little to men and women who know how to take good edible fish—tuna, dolphin, snapper, yellowtail, blue runner, and many others—from the sea and the rivers. They know, too, that they can walk along the beaches in search of turtles that come to lay their eggs in the sand. Turtle egg pancakes are a great delicacy, and turtle meat is said by many to combine the taste of chicken and beef.

From *Living by the Land,* pp. 117-122

6 Tropical Subsistence Homestead

... Any man who converts a rough rocky piece of land into a self-supportive homesite adds in more ways than one to our national wealth and welfare. Rehabilitation of the Floridan Keys, p. 24

Movement back to the land is altogether a practical scheme, since despite our growth in population we still have sufficient space for the purpose. In places where good agricultural land is scarce and population dense, homes and sustenance can be won from forest lands so rugged that they are classed by some agriculturalists as marginal. Marginal means land that is unfit for ordinary agriculture. But no land is marginal if rich in decomposing humus and planted with the right kind of trees. Any soil that has the humus is good, provided it is neither too dry nor too wet to support crops. Too much fuss has been raised about soil quality and fertilizer. Often it is the man who is marginal, and not the land!

Beginning a Forest Farmstead

Living among the trees starts with the building of a simple shelter. Later on this can be replaced by a more permanent and comfortable home. Gradually the woodsman removes the undesirable members of the forest family near his homestead, substitutes productive trees and other plants that he needs, and soon he has a forest settlement that will afford him the basic comforts most men desire.

You will discover that beginning a forest farmstead is not at all expensive. The cost of the land itself is low, the outlay for equipment small. You need buy no work animals or costly reapers, threshers, binders, or other machinery. Your tools? Merely an ax, saw, machete, dibble, and spade.

You need not strip the land to make way for agricultural crops. You need not blast, scarify, or plow. Any necessary clearing and planting can be done gradually. Leave good trees wherever possible. The covering of their branches and the shade they throw make a kind of natural slat-house protecting the crops that grow beneath. Conserve plants that are soil-builders; remove plants and trees that are in no way useful and replace them with the kinds that will help you win a livelihood from the land. Remove the slash to reduce the fire hazard. Cut the tree stumps close to the ground and leave them to rot. Gather loose rocks and pile them around the trunks of trees, where they will keep the roots moist and cool. Conserve rocky hillsides, brooks, little lakes, and beds of wild flowers for beauty. If gullies have been formed by erosion, use brush dams to anchor the soil; if drainage is needed, cut ditches in the ground.

The forest farmstead should be stocked with bees and poultry, both originally inhabitants of the woods. The woodsman can live on tropical tree crops and continuous vegetable crops. Meat and milk are primarily northern cold-weather foods, scarcely essential in our diet. The most cultured people of the East, it is said, depended on figs, dates, grapes, melons, walnuts, and almonds. In the Mediterranean Region the people of the past subsisted mainly on tree crops. Apparently they suffered no ill-effects inasmuch as they developed, in their Golden Age, a great civilization which spread westward and which still constitutes what is best in many of us.

Produce at Home—Use at Home

The heart and soul of any nation is not big business. Big business devours little business and in time becomes international and stronger than the state which nurtured it. The real heart of a democratic country is small businessmen and small farmers engaged in producing and processing useful goods by the proper use of home resources.

The small farm is the basic unit in society and the prosperity and strength of any country can be measured by the number of small, self-supporting homesites which it contains. Thus, the best nations of the world are not those with the greatest natural resources but those with the largest number of little, self-supporting, free-of-debt homesteads. This idea is not new. It is traditional with the people of northwestern Europe, from whom our Anglo-Saxon culture sprang.

Returning to the simple life close to the land, it seems to me, is the only permanent way out of the difficulties that beset the world. As a matter of fact, I offer the subsistence homestead as a means of preserving peace. Settle as many people as possible on homesteads, home-owned and as free as possible of taxation.* The more the farmer can produce on these home acres for his own use, the closer we shall come to the ideal state and the less danger there will be of internal strife. One of the main functions of the state—to provide contentment for its people—can never be achieved unless farms are privately owned and unless their owners are amply insured against various emergencies that are likely to occur at any time.

Errors and accidents in subsistence farming can be compensated for by low-cost, state-sponsored insurance. The group subscribing to the plan could easily absorb the burden of loss from the person who has made serious mistakes in planning or who, through no fault of his own, has bad luck. Personal initiative would not be destroyed by this co-operative plan, but the farmer's life would be more secure and freer from worry than now.

Security—symbolized by a home and an adequate livelihood—is the aim of the masses. And if most of our rural people are self-sufficient, the nation will be. We can achieve self-sufficiency if we first produce at home the things we need, and then sell the surplus, if there is one. Planning primarily for home-consumption, we should avoid putting all our efforts into single, money-making crops such as tobacco, sugar

*With an increase of self-subsistence homesteads more and more people will rise out of the dependent classes. As the need for public relief of these people decreases, taxes could be reduced, since the operation of vast governmental relief agencies has been a gigantic drain of the public treasury.

cane, or cotton. The subsistence farmer can get along with little money. Furthermore, if one crop fails he can get by with the others he has planted because a bad year for one thing may be a good year for something else.

In aiming at self-sufficiency the new agrarian movement deprives the middleman of profits and the silent patent-owner of royalties, but it is in line with the desirable movement to live at home, work at home, and use home-grown products at home. Its aim is security and independence, not cash and luxuries. Capitalistic, profit-making farming is worlds apart from subsistence farming.

The only hope for independence and self-reliance for a majority of our people lies in the small, self-sustaining farm units scattered everywhere over the land. Through consolidation and exploitation we have developed into a land of lords and serfs. I see real danger in a continuation of the old system. I fully believe that every nation will suffer internal friction among its classes unless most of the people have a food supply absolutely under their own control. National prosperity rises and falls in proportion to the size and number of homesteads.

Corporations and Absentee Ownership

American history of the Caribbean Region is filled with the exciting story of the evils of vast estates absentee-owned by gargantuan corporations demanding fat dividends.* Permanent peace and amicable living will never come—*can* never come—until the land is broken into small parcels owned by natives and planted with productive trees.

One of our worst troubles, I maintain, is not race hatred or religious hatred, but class hatred—the arrogance of the man who rides and the humility of the man who walks. Money produces power and its offspring, dominance; but the sublimation of the spirit of dominance is the essence of democracy. *The only cure for the evils of dominance is to control, through legal means, the concentrated power of the wealthy combinations.* If a redistribution of great wealth is needed it

---

*In the tropics many old plantations were bought by big sugar corporations. These were merged into one big plantation, in the center of which stood a sugar mill. The old plantations lost their identity and their ancient charm, and in many instances the former owners became servile hirelings.

can be effectively controlled only by dividing the land into small sections with many owners, because land is the ultimate source of all wealth.

The little man who has his own debt-free food supply is really a big man—free regardless of the accident of birth which determines the pigmentation of his skin and the beliefs he holds. So long as this humble, simple man has land enough to cultivate, he can remain free. Where land is plentiful and cheap there is no problem. When population increases and fertility decreases, as has happened many times, revolution results.

The farm should be a home where a man first raises what he needs for himself and then sells the remainder, right by the roadside if possible. A good farmer sells what he cannot eat; a poor farmer eats what he cannot sell. The farm should be a concrete representation of the owner's personality, the product of his dreams. It is a marvelous accomplishment if he can win a comfortable livelihood from it. A man who can do this is truly a successful man, important to the welfare of the whole nation.

As surely as day follows night, true democracy will mean the death of class hatreds. The small homestead movement carries us closer to this ideal, and it is the only movement that can erase the distinctions between lord and serf. The state should own or control whatever the private person cannot rightfully use or properly operate. Over and above the areas necessary for subsistence homesteads, the state should control the balance for forests and any other conservation projects which would benefit all of us. The proper care of these state lands will furnish work for many thousands of people. After a lifetime of study and reflection I am convinced that the best minds of all ages have held in common the belief that the future prosperity and independence of men are deep-rooted in the soil and in no other place.

When the products of the land are shipped hundreds of miles to immense factories to be processed and then shipped back to the consumer, transportation, processing, and marketing often cost more than the product is worth. Thousands of people are thus uselessly employed.* Actually, there is no good reason why raw materials cannot be processed in local, home-owned factories without being worked by

*See Stuart Chase's *The Tragedy of Waste.*

great corporations which are never as practical, economical, or human-itarian as their apologists would have us believe.

This brings us back to small, local industry and decentralization. The big fish eat the little fish, the big farms swallow the little farms, and big businesses devour little businesses. To get off our precarious perch we must gradually descend the same wobbly ladder we once joyfully climbed, descend to the very bottom, to the small homestead and the small factory in the small community. Big corporations will object, surely, but no other way promises real freedom to the mass of our people. Decentralization and diversification in farming are the necessities of the hour.

In the tropics the first step in the right direction is to plant diversi-fied tree crops in forest formation on subsistence homesteads of at least one acre per person, about five acres per family. *

A five-acre unit measures three hundred and thirty feet by six hundred and sixty feet. In the center of this plot a circle with a radius of about one hundred and fifty feet should be reserved for the grow-ing of tender crops. The remaining land should be planted in close forest formation with hardy tree crops. This scheme could be applied in all climates, with appropriate variations in kinds of plants. A man should have as many five-acre plots as he can successfully handle.

This is the patio system of cultivation, a modification of the old-time *milpah* which the Indians used in Latin America. Milpah is the ancient Mexican name for cornfield. The native cleared a small patch in the jungle by girdling the trees with his stone or copper ax. Then he fired the deadening. His first crop of corn was good, his second fair, his third poor. When the land was exhausted he cut a new hole in the jungle and in time moved from that to still another. As long as the population was small in proportion to the acreage, all was well. The people had sufficient food, but eventually most of the forest was cut and the cultivated land overworked. Rank grasses formed savannahs. Fires swept across the grasslands. The ruined acres were abandoned to the will of nature. The milpah system is satisfactory until it is over-done. But the results are destructive when too large an area and too many areas are cleared.

From *Living by the Land,* pp. 24-25, 27-31

*Five acres in the American tropics is equal to many times that area in the North because growth is continuous and vigorous.

7 Bungalow Construction in Southern Florida

By building low of rock and timber and by giving the main lines of the structure the right proportions and sharp outlines to produce contrast, the house appears to grow out of the land and when surrounded by vines and shrubbery becomes in fact part and parcel of it.

The Everglades and Southern Florida, p. 84

Since coming to Florida, almost ten years ago, I have been designing and building bungalows. During this period there has hardly been a time when I have not been altering an old one or planning or building a new. All the while I have been striving to produce something perfectly adapted to the environment. Long before I could finish one I would discover changes that would cheapen the cost of construction or add beauty or comfort to the structure. I disregarded all precedent, had difficulties with mechanics who would persistently do things the old way until finally I found myself doing most of the work with the help of a couple of negroes, who were willing workers but who could neither see straight nor saw straight.

In this part of Florida we sometimes begin at the beginning by cutting the trees and hauling the logs to the mill. The soil is lime rock, some of it loose, but much of it solid. This is good building material and by blasting, a lot of it may be secured on a small space for house walls, fence walls and roads in the process of clearing the land. The holes when filled with trash and rakings are fine for bananas and pawpaws. By building a kiln of wood and the proper kind of rock a fairly good quality of lime may be secured at a very low figure. With wood, stone, lime, sand and water all off the very lot you are building on, the house becomes in truth a product of the land.

The next step is to buy a galvanized iron pipe and a cheap pitcher pump. A twenty-foot length of pipe and sometimes much less is ample. A coupling is put on the end of the pipe. One edge of this coupling is filed or pounded sharp and opened over the beak of an anvil for a cutting surface. By churning this pipe up and down through the soft, white rock with the help of a little water two men in a few hours can have a pump in good working order—pump, pipe and labor not costing more than a ten-dollar bill.

A pile of planed lumber, costing about $22 per thousand, a case of dynamite, with caps and fuse, and with plenty of lime and water, all is in readiness for business. I find it pays to mix some cement in the mortar and cement is now so cheap that the increase in cost is slight. The center of a thick lime-mortar wall does not harden for a long time. A little cement therefore helps to stiffen it. By building low of rock and timber and by giving the main lines of the structure the right proportions and sharp outlines to produce contrast, the house appears to grow out of the land and when surrounded by vines and shrubbery becomes in fact part and parcel of it.

The natural conditions to be considered are long, dry periods, continuous sunshine for months, very heavy rains and strong winds at times, which drive water in a fine spray through the smallest chink.

This calls for tight, cool, solid, low structures. I should add also that the well water is hard and cisterns are necessary, so that the roof must be of a material that will not taint or discolor or render impure the water.

Although a forester by profession, I do not believe that the earth rotates upon a wooden axis, and I realize also that wood has been used in the past for many purposes merely because of its abundance and cheapness. It is, however, in the end an expensive constructive material if we consider the cost of paint and repairs, the danger from fire and the tribute we pay to fire and insurance companies.

The appearance of it is, however, good and although rock in this section is cheap at the start, even considering the low price of lumber, many prefer the effects gained by a combination of both.

I have used cement blocks, concrete, paper roofing, corrugated iron, shingles, tile, etc. I have even used old barrel staves, cut in half, for shingles. When one lives near the shore there is a possibility of collecting a lot of valuable drift lumber. I have captured ash, mahogany, and Spanish cedar logs adrift in the bay. The tile in my hearth came from the floor of the engine room of a wrecked steamer. The wrecks often yield brass hinges, etc., which are difficult to get in any other way. The enterprising beachcomber can usually find many useful articles along the shore and the waste of lumber on the beaches is enormous since it is seen riddled with holes and rendered useless by borers of various kinds.

Since the roof is half the building, let me dispose of it first. Paper roofing or felt roofing is not very durable, it taints the water and looks cheap at best. Few people desire it as a permanent roof cover, although if carefully put on and frequently painted, it is tight and lasts longer than one would expect under the trying conditions of the tropics.

We have no snow, of course, and steep roofs are therefore unnecessary; in fact the roofs I have built have grown flatter until I have now reached the flat roof stage. A flat roof is easier to build, requires less material and in heavy rains and high winds much of the water blows off instead of into the house.

Shingles taint the water, curl up and open up in the hot sun so that the rain beats in and insects find a fine harbor under them. Corrugated iron is hot and noisy, although extensively used everywhere in the tropics, because it is cheap and quickly put on. It is tight and yields good water. Covered with concrete it forms a fine roof. Tiles are beautiful and cool, but they are seldom tight and since they are usually elevated on strips a couple of inches above the boards of the roof they form a fine harbor for rats and other vermin. If every crack is cemented an enterprising tropical rat will work at a tile till he loosens it. In time he will succeed in pulling out cement enough to squeeze through. Then he has lovely quarters. He could not be safer from intrusion.

I no longer build large houses. I have adopted instead the unit

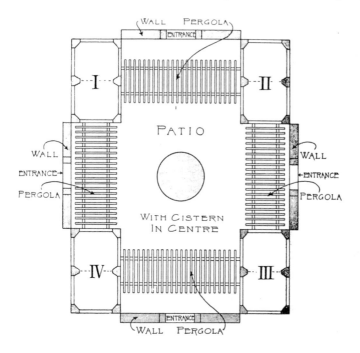

system on the bookcase plan. Each unit measures twelve by twenty-two or thereabouts. These can be built around a central court in any number to suit the size of your family, your lot and your bank account. These may be connected by "blowways" or "dog trots" or "pergolas" or "galleries" or "porches." I was working toward this plan when I struck the following in an article on Chinese art in the International Encyclopedia: "A Chinaman's house, if he is a rich man, is a group of small one-story buildings interspersed with gardens, all within a bounding wall."

That fills my bill exactly, and I am neither Chinese nor rich. The cost of a unit is about $200 and each unit ought to be rentable almost anywhere at $5 per month. Suppose one owns only a small lot. Place a unit on each corner. Connect the units with pergolas and close the spaces open to the street with an attractive wall. In the center one would have a spacious patio.

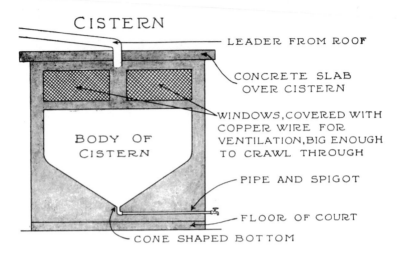

CISTERN

LEADER FROM ROOF

CONCRETE SLAB
OVER CISTERN

WINDOWS, COVERED WITH
COPPER WIRE FOR
VENTILATION, BIG ENOUGH
TO CRAWL THROUGH

BODY OF
CISTERN

PIPE AND SPIGOT

FLOOR OF COURT

CONE SHAPED BOTTOM

In the patio is the place for the cistern, which should be built above ground. If above ground the water may be completely drawn off at any time by means of a spigot. The bottom of the cistern should be cone-shaped, with the apex down, from which the pipe leading to the spigot should start. In that way every speck of sediment may be drawn off at any time.

In the tropics the cistern should be screened and well ventilated. It is cooler above ground than below it. Pump water is always warm in cool weather. If the cistern material is slightly porous all the better. The evaporation will cool the water like a Spanish olla and on the basis of the iceless refrigerator. It is necessary to screen out the mosquitoes since cisterns are their favorite breeding places.

The flat roofs are fine places for solar heaters. A flat tank on the roof into which water may be pumped by hand with a small force pump in a sunshiny climate yields fine, warm water for bathing if covered with glass sash.

The following is a brief description of how I build a unit house. I lay up a narrow wall of rough stone (12x22 feet), a foot or more above the ground. I usually build against boards and pile in mortar and rock. This enclosure I fill with rock, which is packed and pounded down solid. Over the surface of this I lay a cement floor.

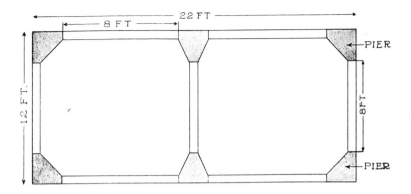

On the cement floor I set up frames of 2x6-inch stuff, each frame 8x8 feet, two frames on each side and one at each end. This leaves room for three piers on each side. These piers are triangular in shape, showing two feet on each face on the outside. They are constructed of concrete, one part cement, two sand and four blasted rock. This mixture is thrown in a wet state inside of rough pier forms.

By making these piers triangular they are strong; it gives a fine space inside for hanging a mirror or picture or for shelves and it avoids sharp corners in the house. The tops of the 8x8 frames serve as a plate on which the roof beams rest. They rest also on the tops of the piers.

All roofs in the tropics should have a good overhang. In early times on this coast houses were built with practically no eaves. They saved

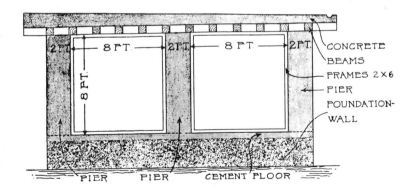

lumber and felt safer in times of storm. Eaves throw the water from the house and shade the walls, thus rendering the house much cooler, since the secret of keeping cool in the tropics is keeping in the shade and in good ventilation.

On top of the roof-beams I lay corrugated iron. Boards may be used instead between the beams and afterwards removed. On this I lay four inches of concrete reinforced with poultry fencing, barbed wire or common galvanized wire of any kind. A rim of cement serves for a gutter and the slope is left to one corner or to the middle of one side. Thus iron gutters are dispensed with. This roof forms a pleasant mirador and a second story may be put on in the same way if the owner desires.

The main part is complete—the finish is easy. A Tropical house should have many openings so as to be all-porch in hot weather and yet tight as a drum in times of storm. Tongue and groove stuff shingled on the outside is good. I use narrow shingles (three-inch) and put one nail in each shingle. A small shingle when it contracts makes a smaller crack than a wider one and if only one nail is used it is less apt to split in the process of expansion and contraction. I prefer shingles and up-and-down boarding to clapboards, since then the rain drips or runs down with the grain of the wood. Good copper screening is necessary, but glass is often dispensed with, solid board shutters being often used.

Such a building is cool and cheap. It has no large timbers in it. It is anchored to the ground by stone pillars and a solid slab of a roof. One of the corner piers may be made hollow for a chimney, and a fireplace is pleasant since there comes a time in almost all tropical countries when a fireplace fire is grateful.

Such a house looks plain and solid—Assyrian or Zuni-like in char-

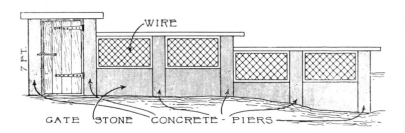

GATE   STONE   CONCRETE - PIERS

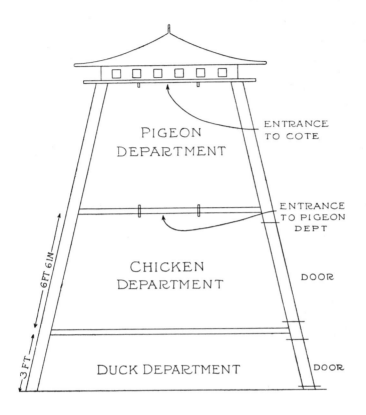

acter—quite in contrast to many of our ornate, gingerbread carpenter-esque constructions, but the shrubbery in the patio and the vine-covered pergolas and fences with many shades of leaf and flower give it all variety necessary. These units may be connected with a fence and the following I have found to be very good and not very expensive: Put up posts ten or twelve feet apart, five or six feet high and one foot square, built in a form of the same kind of concrete mentioned above. Connect these with a wall two or three feet high. Run a 4x4 railing along the top of the posts and fill the space with poultry wire. This is "horse high, pig tight and bull strong," and is at the same time attractive and fine for vines. These unit houses cannot properly be called bungalows, since a bungalow is supposed to be a low, flat, rambling, wooden structure, often with a thatched roof in the East Indies, but the term in America now covers a multitude of sins. One of these unit houses I have built for a garage, but prefer to call it an "autola." One unit may be used for a kitchen and lavatory. In case the baby is cross or some one snores it is easy to relegate them to the units in the farthest corner of the patio. In conclusion let me add that no place, however small, is complete without a place for animals of various kinds, the houses for which may be built in the same way around a central court. Then, too, many people are fond of pigeons. I have built a dove cote twenty feet in the air on top of four posts put slantwise in the ground. Two feet from the ground I have built a floor of boards which serves as a roof for the ducks and a floor for the hens. Six or eight feet higher up I have built another board floor, which serves as a roof for the chickens and floor space for pigeons. The whole is enclosed in netting. The pigeon house has a hole in the center underneath so that they can enter their department from below and thus be safe from intruding hawks.

From *Everglades and Southern Florida*, pp. 83-93

8 Some Interesting Tropical Trees

No man can starve where the cocopalm grows.
Reclamation of the Everglades with Trees, p. 57

The Cocopalm

South Florida is the only part of the mainland of the United States of North America where the cocopalm grows naturally.

The most useful tree of all the world is this cocopalm because it feeds the largest number. People use it for food in some form almost daily and tons of its oil are consumed every year in the North Temperate Zone in the form of soap, hair-oil, butter, and other articles. It is favored as a shampoo wherever hair is washed. The word "coco" is said to come from the Portuguese, meaning "monkey face," because of the fancied resemblance of the eyes of the husked nut to the face of a monkey.

The cocopalm is not found in the wild state; its original home is unknown. The Tropics of the world is now its home; it encircles the globe. The Tropics cover a central band around the earth three thousand miles wide and twenty thousand miles in girth. The Glades are just inside the northern edge of it. A large portion of this territory in the limits of the so-called Torrid Zone is elevated and constantly cold. The cocopalm sticks to the part of it which is alkaline and frost free. The cocopalm is just as much at home here as elsewhere and there are some evidences that the place of its origin might have been somewhere in the Caribbean Region, not far from Panama. It was, no doubt, freely distributed along the coasts of tropical countries to furnish oil for the hair and for lamps and lighthouses. The grove from which Palm Beach gets its name was started from the wreck of a Spanish ship with a cargo of nuts in 1879. South Florida is the only part of the mainland of the United States where it grows and on the Pacific Coast, one must travel fully five hundred miles down the Mexican Shore to find it.

From the standpoint of utility, it leads the procession in the plant world. One must live with it a long time to fully know it, but to most newcomers it has a rugged exotic beauty that attracts and fascinates. With its glistening leaves in the moonlight, silhouetted against the sky or reflected in the sea, it typifies the Tropics. A coral strand or ocean atoll looks deserted, incomplete and sterile without it. It does not demand the sea or salt, but thrives in rich, moist alkaline soil. Some of the best cocopalm groves in the world are nowhere near the sea or salt. Old sailors want to be buried in the shade of a cocopalm by the seashore, and rightly so, since this beneficent tree has, no doubt, saved many shipwrecked mariners from death by hunger and thirst. No doubt, there are many cocopalms that shade the bones of buccaneers and pirates on tropic shores.

Unlike the aristocratic royal, the cocopalm is plebian. It leans toward the water, probably because the soil has washed away on the seaside, causing the roots to be fewer and weaker. It is thus easy to climb and has ridges where the massive leaves have been shed, which give to the barefooted native a good toehold. It is a fitting shade to the hut of a fisherman, for with his sponge-hook he can pull down at any time, a green nut and garner without cost a cool, sweet, fresh, invogorating, nourishing drink from Nature's own distillery. This liquid is under pressure and squirts like soda-pop when the nut is in the soft stage, and is a favorite food with persons who have stomach troubles. It often succeeds when other foods fail. When a child is born in the South Seas, a few more cocopalms and breadfruit are planted and thus the fear of starvation, the worst of wolves, is kept forever

from the door. When the mother's milk fails in the South Sea Islands, the babies are reared on the jelly of young coconuts.

The white meat of the ripe nut is the copra of commerce used in a hundred or more ways and when ground very fine, forms a cream which is delicious on fruits. Coconut-snow is a rich breakfast food. When mixed with fruit juices, it forms a balanced ration. Ice cream flavored with fresh coconut is popular, and rightly so, because it is a delicious and nutritious food and is quite the equal of the other two great tropical flavorings, vanilla and chocolate. In many parts of the Tropics you can see chickens, goats, dogs, pigs and children all feeding in the same yard at the same time on the white meat of the coconut. The pork of pigs fed on coconut is of superior quality and flavor. Set the nuts in beds slant-wise, eyes up, but not too deep. Cover them with trash, rotten seaweed is good, keep them moist and they will soon sprout.

It has been pictured in the past as a perfect adaptation to the seashore. The oily nut floats high and long in water. The husk protects it from breaking when it falls on the hard sand or rocky beach. Soon the little palm springs from one of the three eyes in the end of the nut. The leaves are at first simple, but later form into great compound fronds fifteen or more feet in length. In a few years, seldom under five, it produces bunches of nuts in all stages for many years. Then the mangrove island becomes fit for some smoky-colored, semi-nude sea-islander who from the palm, can garner the necessities of life with a few of the luxuries on the side. With a homemade guitar and a home-made cigar and a homemade hammock of sisal, he can rest in bliss in the shade of the palms. With the fish in the sea and the turtles that lay on the beach, starvation is not possible. Man's ultimate wants are shelter, food, and drink. The cocopalm supplies them all and then some.

In parts of the Tropics the flower stalks are cut while green and tender and to the stub of the stem is attached a light bamboo trough. Several of these may be thus treated and several troughs thus led to gourds or calabashes awaiting the liquid which oozes out and trickles down in the form of a snappy cider or toddy. Think of a bungalow closely surrounded by cocopalms with bamboo conduits leading this cidery juice slowly but continuously into a pitcher on the kitchen table! This palm is the source of sugar in the Far East, also of alco-hol. The strong drink "arrack" is supposed to be concocted from palm-sugar. Fresh palm sugar, called "jaggery," is quite as good and quite as full of flavor as the famous product of the maple tree.

From the outside of the nut comes the husk, from the fiber of which cordage and coco-matting are manufactured. This husk is hard to remove unless you know how, but there is a crab in the South Seas that has learned the trick. This matting has long been preferred for church aisles and office floors. Rope from this fiber does not deterio-rate quickly when wet. It can be pressed into a board resembling leather. The hard shells are often beautifully carved and used for utensils of various kinds. On the shells of these nuts tribal records are often exquisitely and accurately carved.

The wood called porcupine-wood, because of its spiny appearance, is hard on tools and of little value. Like the rest of the palms, it has only one terminal bud. When this is killed by accident or disease, the tree dies.

There are those who grow very fond of cocopalms when grouped by the shimmering, particolored sea of the Tropics with many things, dear to the heart of the native, such as sapodillas, guavas, limes, pineapples, soursops, starapples and other fruits, growing in their protective shade.

The cocopalm and other trees such as the mahogany, mango, avocado, and breadfruit, love the coral islands and so do the natives of the South Seas who carry rich dirt from volcanic regions to cover the jagged limestone in their gardens of everbearing plants. Cattle eat the leaves and the husks are fine for fuel.

Binding the leaves around the stem there is a natural cloth which has been used for many purposes. This may have suggested the weaving of cloth to primitive man. There are hundreds of little islands near to us where the cocopalm and breadfruit grow, the two greatest food gifts of Nature to man.

From *Reclamation of the Everglades with Trees,* pp. 57-62

It is never safe to spurn the common things—they are common because they can endure; they are common because they are fit.

Reclamation of the Everglades with Trees, p. 28

Cypress

Cypress is advertised to the world as the "wood eternal." The largest tree now standing in this State is called the Sovereign Cypress, near Sanford. It is seventeen feet in diameter, almost a rival of its famous relative the Redwood of California. The old world cypress is an emblem of mourning and sadness and cypress trees the world over have been planted in cemeteries.

Our native cypress drops its leaves for a short time in winter; its trunk is a silvery grey; it is often heavily festooned with epiphytes or air-plants, and in groups in sloughs and prairies form striking features of the landscape, especially when they begin to break into masses of pale-green foliage. Aside from the superior quality of its wood and its beauty as a typical part of the Southern landscape there is apparently no place too wet if its roots can get a foot-hold in the mud. It grows in places, even in lakes, where the water may be several feet in depth for months at a time. The old trunks swell into abnormal shapes due no doubt to the irritation caused by water. In swamps the cypress as well as some other trees produce upright branches from horizontal roots called "knees." They are supposed to furnish air to the roots,

possibly help support the tree or for some other purpose or no purpose at all. On the highland where the tree also grows these knees fail
to form. They have something therefore to do with the water and
mud in which the tree grows. Anyway, this noble tree has been striving to cover the Everglades for ages. Before the white man the Indians
used it for canoes. It would have covered this mucky land long ago
but the odds were against it. It can fight water and wind but falls
before the axe and fire.

Pine

What the cypress has been to the lowlands the old Caribbean-pine
has been to the sand and rock-lands, although it too will grow in
swamps in case the water is not too deep or long-standing.

The Caribbean-pine is in some parts of Florida called slash or
swamp-pine. In Southern parlance a slash is a wet place overgrown
with bushes. The specific name *caribaea* is spelled with one "b." I
presume this was the original spelling and holds by the law of precedence. There could be no better name than the Caribbean-pine. The
word "caribe" commonly used in the West Indies means very wild.

This tree grows in Florida on rock and in sand, down on Big Pine Key, over in the Bahamas, in Cuba, the Isle of Pines and on upland dry ridges in Central America. It grows sometimes in swampy places and at one time, according to Chapman, grew on the Island of Key West. It is a tropical pine and in many places grows beside the palms and mahogany.

One of the many striking pictures of the Florida of old, now rapidly passing, is the old gnarled Caribbean-pines picturesquely silhouetted against the sky. One must admire their sturdiness, bending and twisting in the wind with the storm clouds sweeping by during a tropical gale.

Although the Caribbean-pine is a tough old tree they will soon be gone and be replaced by another landscape with other trees from distant places. The wood produced by this old tree on hot dry rocks is also tough, heavy and hard. It is never safe to spurn the common things—they are common because they can endure; they are common because they are fit. This old pine lumber is used when green. Even then it is necessary to soap the nails to get them in but once in they hold for keeps and many of the old houses built of this wood are sound and still standing. We must not forget that it is the wood that grips the nail and not the nail the wood. The trees were cut and burnt to give way to groves of fruit. Log rollings were common. Sometimes they were used to burn out the stumps, sometimes for lime kilns. In the construction of rockhouses, the rock, the wood, the lime, the sand and the water were all from the same lot.

Over a period of many years this old pine has been scorned and wasted by lumbermen and carpenters. Thousands of cords are burnt every winter in fireplaces. Old stumps, roots and knots full of pitch yield a bright, hot and lasting fire. Although deadenings, where trees were girdled to die standing, are now not common, the heavy heartwood is still in demand for fence posts and firewood. On some exceedingly poor rocky soils it is crooked, twisted and grotesque; in other places where it gets what it wants it grows thick and tall. In choosing land it is well to look up as well as down. When the tree gets big and the top heavy a hurricane may topple it over, but when it goes tons of rock clinging to its tough roots go with it. It is a good windbreak for any country where it can be grown. It will do its part in breaking the fury of the gale.

From *Reclamation of the Everglades with Trees,* pp. 26-29

Many of nature's experiments have been failures, many still existing are failing. The banyan experiment looks like a success.
Reclamation of the Everglades with Trees, p. 36

Native Rubbers

There are few trees of more interest than the many odd species belonging to the fig or rubber family. The world has known the East Indian rubber for years. As a pot plant it was in every vestibule. The leaves were wiped clean of dust by every housewife. The rubber in the hallway was usually attended with the same scrupulous care as the canary or the cat. It yielded rubber in India in the early days, but better trees of another kind were found in Brazil.

Our great interest in the banyan is botanical. It forms a forest in itself—a forest that is not easily uprooted and a tree which really never dies unless killed by accident. The true banyan is called *Ficus banian.* The common East Indian rubber is *Ficus elastica.* The leaves of several species are cut in great quantities for camel and elephant fodder.

The Hindu word for trader is banyan, and the merchants no doubt spread their wares in the shade of these trees for sale. They were probably the first market places.

The largest banyan tree that I know of is on one of the Howe Islands in the Indian Ocean. It covered nearly seven acres. One near Calcutta lost a couple of acres in a hurricane. In referring to the banyan the poet Milton quotes from Pliny the following words:

"Branching so broad along that in the ground the bending twigs take root; And daughters grow about the mother tree, a pillared shade, High over-arched, with echoing walks between."

These trees have really reached a high state of development. First they start as an epiphyte high above the ground where they are safer from molestation. They seek the shelter and protection of another tree. An epiphyte is a plant which lives upon another plant without stealing any of the sap from the living tissue. It prefers a hollow place where the tree is injured—a place where a limb has been broken or in a crotch where moisture and humus have accumulated. The hollow at the base of the fronds of a palm is a favorite starting place. The seed is carried by birds. These rubbers therefore have developed the habit of starting above ground and use birds for distribution.

The roots of these figs are delicate creeping tendrils at first. Like tender threads they descend to the earth where they gain a footing. Then the tree becomes a terrestial plant and is no longer an epiphyte. These air-roots soon grow thicker and stronger. They envelop the tree trunk. They anastomose or grow together and the tree which is now a host is strangled to death. The roots, now stems, graft together. The dead host rots, the strangler feeds upon its decaying body which is soon consumed. The roots have become trunks and finally one trunk. Then the tree trunk begins to form rings of wood like other trees but a section of the tree will show several small trunks all joined in one or encased in another trunk. This strangler-fig habit is exaggerated in the banyan. The roots drop down from the limbs and the tree in time becomes a forest in itself.

This banyan habit seems to me like a very high state of plant development. Although the flowers of this tree are insignificant they are usually fertilized by insects of some kind and birds are utilized to carry the seed. Strangest of all is this development of longevity, because the tree or parts of the tree can live on forever unless killed by accident. It can usually withstand typhoon or hurricane. It is a natural slat-house. The tree produces forest effects. Instead of many trees growing together crowding one another this tree possesses the land and joins the whole into one congenial formation. If the main trunk is not in a fertile spot some of the minor trunks better located will in time supercede it and thus feed the whole. Any one of these trunks separated from the rest would go on growing. It is usually rated as a weed—a strangler. It is really a cooperative colony. They are cut with machetes by foresters and checked in their strangling habit but if a use could be found for their wood they ought or might prove the strongest and best of all our trees. They all contain rubber but not enough for commercial purposes. This rubber, however, was put there no doubt by nature to protect the tree from injury. It forms a scab over wounds. It is also fireproof. The soggy wood when green refuses to burn.

There is a lesson worthy of careful study in the banyan. The way to learn the right way is to study nature's way. She has been experimenting a long time. Plants in man's hands lose their personal initiative. Many of nature's experiments have been failures, many still existing are failing. The banyan experiment looks like a success.

In Florida there are two native species of fig, *Ficus aurea,* the strangler-fig and *Ficus brevifolia,* the poplar-leaved fig. Out in the

Everglades region, wherever the land is a few inches above the general level, these wild rubbers grow. On canal banks, in fact every where the seeds can find a lodgment they grow with great vigor.

The roots of some of these rubbers are often of such shape that they penetrate clefts in rocks which in their process of growth, they pry apart or flatten themselves out over the surface of the rock. They accomodate themselves to all kinds of situations.

It would pay us well to study carefully the uses to which the Indians of North and Central America put certain trees. Much of that we have used is already of Indian origin. Careful study of ancient records and archaeological relics might yield some very valuable hints—in fact these Indians were many centuries in one location and their use-knowledge of the things around them was equal to that of today; in fact many of these uses have been neglected and forgotten. In history and archaeology, you cannot separate the plant, the place and the people. They must be studied together.

In tropical America there are many species of Ficus, little known, with their roots in strangling embrace of others resembling the writhing of a mass of serpents or constrictors which live in the tree tops. The Indians called them the snake-fish or eel trees, because in the wetness of the tropical forests, these tortuous slippery roots resemble a mass of wriggling slimy eels. They all yield rubber of some kind. It is in the milk of close to one hundred species of plants, trees, vines and bushes. It is found in small quantities in many species of plants, even in the beautiful goldenrod, so common in northern fields and by northern roadsides*

From *Reclamation of the Everglades with Trees,* pp. 33-38

. . . one of the most remarkable members of the plant world.
Reclamation of the Everglades with Trees, p. 42

Australian Pines or Casuarinas

Although natives of faraway Australia and the South Seas, these trees encircle the tropics of the globe. For binding shifting sand on the shores of the sea in windy situations, they have few, if any, equals.

[*Thomas Edison at his laboratory in Fort Myers, Fla. was investigating goldenrod as a likely source of domestic rubber. Ed.]

They may be trimmed into a windbreak hedge to such a height that homes in its lee will be protected in times of severe storms.

The wood is hard and heavy. A man was moving a brick house at Miami Beach. His gum rollers gave way. He could not get new rollers and the house was stalled in the middle of the street. Nearby were some Australian pine logs. He hastily converted them into rollers. They were as good as gum. The destructive or dry distillation of this wood would probably yield a great variety of products besides the residue of charcoal.

Some trees bend with the wind and let the wind slide over and trickle through. Some, like the willows and gumbo-limbos let their limbs snap off sprouting out afresh from the old stump after the storm is over. The casuarinas have adapted themselves to such conditions since they have no leaves. The green parts are pliant branchlets. The stems are sturdy and give with the wind. They have adjusted themselves to the stormy seashore.

Casuarinas are called *she-oak* in some places and Australian pine in others, but they are neither pine nor oak, although their wood resembles that of oak and in general appearance they are not unlike pines. These trees are also called whistling pine, Polynesian ironwood, beefwood and a host of other native names, but its scientific name, *Casuarina,* is not difficult and is appropriate since it means "with branchlets like the feathers of the cassowary bird." The branchlets are green, pendant, and without noticeable leaves. These trees are favorites of the natives of all sandy tropical shores and are highly praised by almost every forester in tropical regions.

There are many old casuarinas in South Florida. On the seashore of Biscayne Bay, not far from Homestead, there was a group of them called "The Cedars." For sailors these striking trees formed a landmark that was conspicuous for a long distance. They were the leftovers of an old nursery which had obtained the seed from Cuba.

Casuarinas grow on salt marsh, seashore sand, and rock and muck at the rate of ten feet or more a year under favorable conditions and they naturally grow straight into the air. They are long-lived and may reach a height of one hundred and fifty feet. Being gross feeders, they have a very extensive and sturdy root system. To uproot a stump two feet in diameter is a real job for both man and dynamite. They fruit abundantly while very young and are easily injured by fire and cold, but are remarkably free from disease.

The common species is *equisetifolia,* meaning "with jointed leaves like horsetails." There is another species in the Redlands region. I am not sure who first introduced it and am not sure of its specific name, but think it is *C. lepidophloia.* It is a beautiful tree, especially adapted to a rocky limestone soil and characterized by the production of many shoots from its roots. It soon produces a thick barricade and can be easily and quickly propagated from these root suckers. This species is rapidly spreading throughout this region because many people like it better than the common original kind. I believe that this species is dioecious—male and female organs occurring on separate trees—and that we have only the male kind, for our trees produce no seed. I doubt if you can be sure of the identity of a species if you have only one sex and no seeds.

The wood of casuarinas is pink, turning dark with age. It is heavy, hard, and strong and is excellent for fuel and scaffold poles. It produces an enormous amount of wood in a very short time on soils where few other trees will flourish. Like many other tropical woods, it is seasoned with difficulty. The trees should be sawn as soon as cut and carefully piled in a ventilated, shaded place. Small straight poles of casuarina could be quickly grown for various purposes; even, it is said, for paper. This wood is used for war clubs by the natives of the South Seas.

C. equisetifolia appears to do as well here as anywhere in the tropics. This tree seems very much at home on the Keys and for many years has been a favorite as a shade tree in Key West. Although not a native, this tree may be classed as a naturalized immigrant and is so listed by several botanists. According to the botanists this tree has bacterioids on its roots and probably conserved nitrogen and improves the soil as do so many plants of the bean family. Other things however do not grow in its shade and the ground is usually covered with a carpet of fallen leaves. The bark of these trees is rich in tannic acid and excellent for tanning leather.

The Redland species (*lepidophloia*), with many little ones springing from the roots of the mother tree, naturally forms a group capable of withstanding the severest gales. Wind hitting such an incline is diverted upward, forming a protected area of large extent in its lee. When growing in group form, the trunks are protected from the scorching and drying effects of the summer sun.

These trees take full possession of the land and although easily killed by cold and fire withstand almost all other forms of destruction except of course the axe.

These casuarinas may form a very welcome mantle of green if our native Caribbean pine becomes a thing of the past, owing to excessive cutting for lumber and fuel.

Over on the keys of Florida the seashore is bare solid rock washed clean by the waves. Just above normal high tide I bored a hole in the rock with a crowbar the size of a pint funnel. In this I put a seed of the casuarina and lightly covered it with a layer of ground rock. It sprouted and grew and is now a large tree by the seashore often splattered with the salt spume of the ocean.

From *Ten Trustworthy Tropical Trees*, pp. 73-77

Pencil Cedar and Spanish Cedar

Most men would pick a poor cigar from a good box, rather than a good cigar from a tin can.
Reclamation of the Everglades with Trees, p. 53

Centering about Cedar Keys on the Gulf Coast the pencil-cedar grows in swampy places and is worthy of a trial in similar sites farther South. Thoreau was one of our first pencil men. He not only manufactured the pencils but used them to good purpose. In 1845 he built for himself a hut in the woods and lived in solitude. After this he moved into Concord and pursued the trade of his father, a pencil manufacturer. I like to associate this master mind with good lead pencils. A good smooth-working pencil not only marks the contact of thought with paper but the flow is stimulated or retarded by the character of the pencil. Reeds were used for pens, also quills, but some ingenious individual filled a reed with carbon of some kind and thus began the pencil. Graphite, which means "to write," was the kind of carbon used. It is the little wooden cases which hold the graphite with which the forester is especially concerned. The Floridan foresters should be interested because the best pencil wood is produced by the pencil-cedar of Florida. This tree is in danger of extermination because of the constant demand for its wood. It has been reported to me

that there are thickets of young pencil-cedar on the West Coast of Florida that will survive if accorded a reasonable degree of protection. The wood is soft, easily sharpened and fragrant.

The term cedar is loosely applied to almost any kind of wood which is soft, light and fragrant. The term true cedar probably only applies to the genus Cedrus. It is safe to say that the cedar of Bermuda, of the Bahamas, of the Barbadoes and other tropical islands is very much the same as the pencil cedar of Florida. It is not very different from the well-known northern red cedar familiar to all because of its fragrant wood used in closets and chests to resist moths and for pails and other small articles of woodware. The northern red cedar is common in old fields and along fence rows where the seeds have been dropped by birds. In fact there is a bird called the cedar-bird. This probably accounts for its wide distribution. Its tropical sister the pencil cedar is softer and more suited to pencils. There have been many substitutes even of such good wood as the incense-cedar of California, but for some reason or other none wholly as good for pencils as the Floridan cedar. Both pencil cedar and white cedar grow in lands that are wet and mucky, and for excellence among light woods have no superiors. You can hold a wooden pencil safely in your teeth, in your hair, or back of your ear. It must sharpen well and easily in modern sharpeners. Some busy men cannot work well without a full supply of carefully sharpened pencils. Some must have pencils of many colored fillers. There is a homely philosophy in the pencil, and I fancy Thoreau enjoyed the making of pencils because to him it was a necessary tool for the expression on paper of his marvellous interpretations of Nature. Some of our most cherished bits of literature were hastily written on scraps of paper with the stub of a lead-pencil. Some of our artists use pencils of very high grade. Rubber was so called because it was first used to rub out pencil marks and after all one of the great virtues of a pencil is the fact that its mark may be easily erased. It seems desirable that Florida maintain its reputation for the production of high-class pencil-wood. What is good for pencils may be good for many other things and it grows quickly and naturally in our mucky swamp lands; in fact, better here than elsewhere.

Continuous and constant demand for such woods as Spanish-cedar for cigar boxes, pencil-cedar for pencils, even common pine for orange crates, slowly but surely eats into our supply of woods for special purposes. They will be sorely missed some day. Even now various trades are searching for suitable substitutes. In the case of cheap cigars they are using poplar and other white woods and pasteboard on which they print with considerable skill the grain of the true cigar box cedar wood. In some cases cedar oil has been used to give to other woods the fragrance that belongs to cedar. Most oldtime smokers want a cigar box of true Spanish-cedar because it is supposed to repel bugs and increase the fragrance of the cigar. Its virtues are now retained with a foil wrapper or a celophane covering. This probably does little good insofar as the cigarette beetle is concerned since he infests many cigar factories, his eggs are wrapped in the cigar and in time eats his way out. Cedar oil probably has little effect on a bug that chews tobacco. When a cigar is being smoked and there is a slight unexpected sizzle it is probably a beetle that is being exploded by the heat.

Of these living things on earth which are gradually passing under the juggernaut of modern industry, we are only custodians. Use or usufruct does not mean destruction. When a living creature is completely destroyed, like the dodo, it is gone forever. It can never be duplicated except in pictures. All these things have great educational and scientific value, even if they cannot be used in industry.

From *Reclamation of the Everglades with Trees,* pp. 30-31, 51-53

We ought to at least produce our own supply of fish poles. A fishpole
farm ought to prove a profitable novelty.

Reclamation of the Everglades with Trees, p. 69

The Bamboo

The bamboo is usually associated with the Far East, mainly with
the people of Japan and China. After centuries of experience they
have learned to use it in countless ways. It requires the deft hands,
patience, and experience of the Oriental to convert the bamboo into
many articles of daily use.

First, it is a tree-grass, second, it grows with marvelous rapidity,
and third, it possesses great strength in proportion to its weight.

Nature's experiment in a hollow pipe-like stem is successful in the
bamboo. The stems are reinforced at regular intervals with nodes and
partitions. To add to the strength the stem is encased in a layer of
silica or natural glass. In a tropical gale they sway and creak in the
wind and if flattened to the ground send up fresh shoots which in
good soil grow sometimes at the rate of three feet in twenty-four
hours. In fact, in a hot climate and rich soil a bamboo reaches a height
of thirty or more feet in as many days. It has great strength, splits
straight, and so many uses that it is useless to try to name them. They
are cut into canes, strips, splints and shavings. They are used in baskets, pipes for water, hats, mats, house construction, pipe stems, chop
sticks, plant pots, flutes, and other musical instruments, Victrola
needles, pens, broom and brush handles, fish poles, bird cages and all
kinds of furniture, pumps, ropes, paper, and bamboo shoots for food
creamed like asparagus. Over one thousand uses have been listed.
Many kinds of bamboos have been introduced into the southern
United States and so far have served more for ornament than other
uses.

With its feathery plume-like foliage it is surely a thing of beauty
and bamboo groves when cleaned and well kept are a never-ending
curiosity to even those who are accustomed to a tropical environment.
Like the palms they can be used to great advantage in landscape
planting if properly placed. Like many other striking tropical landscape features you must live with it for a long time in order to become
fully attuned to it. The plumed bamboo, the exquisite tree fern and

the patrician palm are exotic to northern eyes till there has been time for friendly acquaintance.

The little hollow culm of grass is a bamboo stem in miniature and a grove of bamboo is merely a giant field of grass grown big under the stimulation of a tropic sun and soil. A grass field in the north and a bamboo grove in the tropics is a fair measure of the difference in productivity of the two zones. It is like the difference between the ferns of the temperate zones and the magnificent tree ferns of the tropics, surviving remnants of the great carboniferous age when these trees corraled the sunshine of long ago in the form of coal.

Like the elephant and water buffalo the bamboo needs the Oriental. Many centuries of use-association have united the two to such extent that I doubt if the bamboo ever becomes popular in the western world. The American is used to the axe, a board and a pocket full of nails. It is not, therefore, only a question of growing a certain kind of tree—the habits and demands of our people must first be considered.

The cane-brakes of our South are small bamboos. These culms are woody and grow to be ten or more feet in height. Southern reeds are also of the same kind only more slender in nature. Every Southern country boy is familiar with the cane-brakes. They are excellent hiding places for wild game and runaways. The cane-brakes are really bamboo thickets and although common and well known to all of us are really of as great botanical interest as is the oriental bamboo. There are two canes in our South, the large-cane and the switch-cane. The large-cane inhabits the alluvial bottoms, more or less submerged throughout the year. The switch-cane occurs on land less subject to overflow. The area occupied by the switch-cane has been reduced in clearing land for agriculture. This dense growth was conquered by fire because in dry times this small cane burns with much crackling. These two canes or bamboos are alike in habit but differ in their mode of reproduction. In the case of the switch-cane which grows to be ten or more feet high in height the slender stem branched from the base is seldom more than half an inch in thickness. These tall canes are flowerless. Once every three or four years early in the spring flowers are produced on naked shoots only eighteen inches high. Although the slender canes spring copiously from the rhizomes or underground stems, the flower or seed-bearing shoot is only produced once in every three or four years.

In the case of the big-cane which reaches a height of twenty-five or more feet and an inch in diameter, panicles of its flowers are produced in the axils of its branches at long and indefinite intervals of time. In certain parts of our South old residents were astonished to suddenly see the large cane-brakes bending under the burden of heavy nutritious grains on which birds and beasts were feeding. This may occur only once in a person's lifetime. The farmers thought a new plant had suddenly appeared in the country. The stock grew fat on the seeds. Quantities were picked and stored for future use. After this crop the plant dies and the cane decays. From the seed a new crop comes. During the first year they are simple sprouts. They furnish sweet and tender pasturage and are called "mutton-cane." In time they grow into a dense thicket reproducing for another half-century from their underground stems, forming the almost impenetrable cane-brakes in which the bear finds the securest retreats. It is strange indeed for a plant to reproduce itself from its underground stems, for a half-century, then bear an immense quantity of seed and then die. Mysteries are not confined to the bamboo groves of the East. They can be found in the common cane-breaks of our own South.

From *Reclamation of the Everglades with Trees,* pp. 63-67

If one plows this soil he must use dynamite, and all weeding is done with a machete or sailor's sheathknife.

Ten Trustworthy Tropical Trees, p. 47

The Lime

Visit a conch farmer on the Florida Keys and the conversation will soon drift to the condition of his "sours and dillies."

The "sours" or limes were planted long ago mainly for their acid juice which was cherished by sea-faring folk to combat scurvy, while "dillies," the short for sapodillas, were grown because they have always been held in high esteem by the natives, both black and white, of the Florida Keys and the Bahama Islands.

The buccaneerish taint in my blood got the upper hand when I bought a farm on the keys well stocked with limes, sapodillas and coco-palms, and a sloop which I named "The Dilly." Since then my interest in sours and dillies has grown, in spite of devasting storms, tricky commission men and long droughts.

These two fruits grow together on the keys among lime rocks of coral origin, where soil is often so scarce that on some acres, which one could easily select without wandering far, a man would have to scrape with a spoon for a whole day to get a barrow load. The rocks stick up as though the bones of Mother Earth were dry and bare, without skin or flesh of any kind.

In the crevices of the rock there is some soil, and from the porous rock itself the plant must derive nourishment. At any rate, the lime tree produces sour limes, and the sapodilla tree sweet sapodillas in great abundance.

If one plows this soil he must use dynamite, and all weeding is done with a machete or a sailor's sheathknife.

In the moist season the little lime, hardly more than a seedling is planted in a rock crevice or pot-hole. If the ocean keeps its place and the weeds are kept in check, the lime tree will thrive and, in three years, will blossom and fruit—a fruit with a delicious refreshing aroma which puts the lemon to shame. The lemon is a coarse, thick-skinned, rough, raggy and acrid product compared with the lime. School children in Boston eat limes pickled in salt water, at recess. The lime is a naturally refined and delicate acid fruit.

The lime is a spiny, semi-wild crop, although a spineless variety from Trinidad is being tried. It stands no frost and will not flourish if too carefully tended. No fertilizer except a little half-rotted seaweed and no cultivation except a couple of weedings a year, are needed.

Heavy crops of fruit are produced almost every summer, often with a light winter crop, and the limes from the keys are especially cherished because, unlike mainland limes, they will carry long distances without deterioration when ripe. The lime is thin-skinned, full of juice in proportion to rag, of a delicate inimitable aroma. Once a lime-convert the epicure forever after spurns the lemon.

My crop last year on about four acres of land amounted to two-hundred and some barrels. A flour barrel is the standard and holds about one hundred and twenty-five dozen limes. They netted me on the average $3.50 a barrel. They probably retailed at twenty cents a dozen costing the consumer about $25.00 a barrel—a fair instance of the abysmal gulf between the consumer and producer.

Lime juice has other uses than assuaging thirst. In the form of citric acid it is extensively used in manufacturing establishments. A little lime juice put in the water in which meat is boiled renders it more tender and palatable. Added to deserts, other fruits, jams, etc. it brings out their peculiar flavors and removes flatness. It offsets hardness in water. With salt it will clean brass and remove stains from the hands. It improves and whitens boiled rice and sago. It is a soothing application to irritations caused by insect bites. It is better than vinegar as a salad dressing. It makes a cleansing toothwash diluted with water. It is good for the liver, useful in fevers, and they say, a little lime in the water you drink is sure death to the typhoid bacillus. And so I manage my lime plantation, on a kind of *laissez-faire* system, but it pays a good interest. A newcomer would hardly notice it in passing. A colored man called Parson Jones, otherwise known as the Sultan of Caesar's Creek, has an eye on it. Every month or so I meet him in town, but his good wife, who picks limes also, has not been away from her island home for three years. Three or four times a year when we want to bathe in the briny parti-colored waters of the keys or seek plunder on beachcombing expeditions along the shores, I drop in to look at my plantation and pick some green coconuts for the refreshing liquid which they contain.

In the Lesser Antilles of the West Indies where the lime is extensively cultivated concentrated lime-juice is bottled. This is specific for the dreaded sea scurvy. With weevily flour and wormy cheese with no fruits or fresh vegetables seamen suffered severely on long voyages. British ships were forced by law to carry it and the old-time square-riggers were known throughout the world as "lime-juicers." Pirates planted lime trees around water holes and springs in the Tropics to have a handy supply. Canova speaks of the great crop of golden limes around the famous Harney's Punchbowl, a spring of water in the rocks now in the city limits of Miami.

I find that water with a little salt and lime juice in it a very useful medicine in the heat of a tropical summer.

These days the limes suffer from drought and too much brackish water in the soil. I have found that piling rocks around the trunks holds the trees in place in times of storm. It keeps the roots cool and encourages the formation of dew. I began this forty years ago and at last it has become an established custom. It shows how long it takes to even get little, very simple and very obvious things under way. I noticed that on still cloudless nights I could hear the dew drip among the loose rocks by getting close to the ground. The beneficent dew is the savior of crops in places where water is precious. In order to form it must have some solid object to form on.

The lime can be used when green. Even half-grown limes contain a lot of juice. If the tree has plenty of rain it will produce big limes. In times of drought they will ripen when no bigger than a hickory nut.

In general however a fruit waves a flag when ready to be picked. It needs no chemist to say when. When ripe the lime is bright yellow and very fragrant. It sometimes hangs in ropes and bunches and the trees are usually free from pests if the surrounding conditions are natural. There should be plenty of birds, lizards, etc., to keep the insects in check.

From *Ten Trustworthy Tropical Trees,* pp. 47-51

The Mango

Were I planting a grove of mangoes again either on pine or hammock land, no matter how rocky, I would gather seeds of many good kinds and dibble them in the ground with a dibble-stick. I would plant them in forest formation close together and then later remove the kinds I did not want. I tried this once on two-and-one-half acres. Gopher turtles ate many of the seeds and my neighbors snitched the seedlings one by one till there was nothing left.

If a northerner wants apple pie in the Tropics he can have it without apples by using green mangoes. If he must have lumber the mango will yield it. There is no reason why a fruit tree cannot yield lumber. Apple wood up north has been used for saw handles and pipes for ages. The walnut yields both wood and nuts. The mango yields a wood something like northern ash and is good for many things, especially boat oars. We can use oars to good advantage fashioned out of native wood rather than import them from northern points of manufacture.

Some call the mango the apple of the Tropics. The novice in eating the old seedling sorts meets with difficulties. Some kinds have the smell and taste of turpentine full of cottony fiber. Some have the aromatic smell of an old-time barber-shop. It is mushy, slippery and hard to hold. The juice is like yellow paint. You must retire to wash and pick the fibers from your teeth. But in the better sorts the seed is small, the aroma pleasant and the delicious peachlike pulp melts in your mouth. In spite of this there are many old timers who eat the old time mangoes, to them a relish acquired over a long period of time.

Some negroes in the Tropics practically quit work to take full advantage of the mango season and, as with apples up north, will eat the fruit long before it has time to ripen. There are mangoes that are tart and mangoes that are sweet, mangoes that can be cooked and mangoes that can be sucked like an orange. There are many kinds. In the Far East, potentates have walled-in gardens where they carefully guard varieties of special merit which they hand down from father to son like heirlooms. I have heard it often said that the natives of some Spanish American counties never start revolutions except in mango season. In that way they can be sure of something to eat.

The mango belongs to the disreputable sumac family, to which also belong the poison-wood, the poison ivy and the famous cashew nut. The mango is poisonous to some people, but the poison is mostly in the stem of the fruit and care should be used to avoid dropping the juice of the stem on a tender skin. Some people are so susceptible that to touch a knife used in cutting the fruit or a towel that has been used in wiping the fruit will cause an irritating rash. I know of one woman who says she is affected by standing in the shade of a mango tree. This seems impossible but I do know that many people are severely afflicted by a mango rash. I know of a colored woman who suffered from a rash which the doctors could not cure. She finally admitted that she had been eating mangoes and that she would rather have the rash than do without mangoes.

The mango is a beautiful broad spreading shade tree. Its rounded crown and dense foliage form a perfect shelter from the sun. It is never leafless. The young leaves are usually a pinkish red. The mango tree usually produces a great profusion of bloom but only an insignificant percentage gets fertilized. I have noticed a branchlet with literally thousands of flowers finally producing only one or two fruits.

I have planted mangoes all my life and always unconsciously associate them with India. A well-known Hindoo quoted a Chinese saying that "every country to be great must have a great tree," and added that *the* tree of India was and is the mango. Although highly developed in Indo-China almost everybody associates it with India. There is nothing sacred or mystical about it although it probably enters more intimately into the life and literature of the Hindoo than any other tree. The Hindoo friend referred to above always visits South Florida in the mango season. He comes for tree-ripened mangoes and not for the climate. I am always thrilled when I hear the musical thud of a tree ripened mango when it falls to the ground in the stillness of the tropical night.

Natives resort to all kinds of tricks to force mangoes to bloom. Dry weather or cool nights will often check vegetative growth and cause bloom. Some however hack the trunk of the trees with machetes and some almost completely girdle the tree.

Some kinds of trees normally fruit every year, some every two years, some only now and then, some tropical trees continuously fruit and some fruit but once and then die. In my opinion, to change this cycle of production weakens the tree and shortens its life span. Mangoes are treated rough in many places giving rise to the old saying in India "dogs, women and the mango tree, the more you beat them the better they be."

I repeat that in case I plant another grove of any kind it will be seedling mangoes. It will be a fruit forest, carefully thinned from time to time. If for any reason it does not succeed for fruit it will still be a forest worthwhile and will be primarily for my own use or for local sale. I see no reason why any of us should support the railroads and a lot of in-betweens. We have been receiving whatever they have been willing to give us and have been content eating and canning the culls.

Mangoes grow as well in South Florida as in any place I have ever been. We have a greater variety of choice kinds than I have ever seen elsewhere and although it may never develop into a great commercial fruit for shipment north it will always be a great favorite with the native for home use. After all this is what we want and if others like the mango and like it enough they will, like my Hindoo friend, come here to eat it in the mango season. It would be far better to bring the people here to eat it than carry it to the people miles away.

From *Ten Trustworthy Tropical Trees,* pp. 25-32

It has been hard to convince people that the genuine, old-time mahogany is native to Florida: that it was in fact, common at one time here.
Ten Trustworthy Tropical Trees, p. 37

Mahogany

Floridian mahogany is now in such great demand for furniture and boat construction that every acre of wild land is being searched for trees. Airplane reconnaissance is now being used to spot the big trees in dense hammock islands in the mangrove swamps.

Unknown to the public there was a spot in South Florida where a group of virgin mahoganies was surrounded by swamp. These trees were three or more feet in diameter, breast high. They were broad-spreading with humus a foot or more in depth beneath. Living in the surface duff were rats, which fed on the seeds of the mahogany and false mastic trees. Rattlesnakes in turn fed on the rats. Lysilomas and fishpoison-trees were also present. These trees have all been cut. Perhaps there are other such spots that might be preserved to show this and coming generations what once was widespread. We fail to realize that the landscape is constantly changing through man's interference, and what some of us saw in the Florida of old will never be seen again.

When I first visited the Florida Keys I was amazed to see the natives cut a magnificent mahogany, carefully select the natural crooks for boat construction and leave the main log to rot in the bush. I have often found logs buried deep in the natural duff of the jungle.

The natural crooks were buried in the mud for several weeks. The bark would loosen from the wood and underneath there would collect a jelly-like mass of stuff which had oozed out of the wood in the process of osmosis. Then they would shape the timbers to meet their needs and lean them upright against a wall in a dry and airy place for months till they were perfectly seasoned. Wood thus treated lasted for a long time in the sturdy boats which they needed for fishing, sponging and wrecking among the Florida Keys and reefs.

There was an old house in Key West. It was painted white and had withstood the storms of many years. Somebody scraped off the thick coat of paint and found it to be solid mahogany underneath. I always whittle off a shaving of painted chairs and in several cases have found rich red mahogany underneath. I once saw a pig pen of mahogany poles and in a thatched native home in British Honduras I saw floors of two-inch mahogany planking worn smooth by the horny feet of several generations of well-to-do black mahogany cutters.

A log of mahogany cut on the island of Andros had a peculiar caterpillar grain. It was sent to England and sold in the auction market in Liverpool. It was bought by an American furniture firm and converted into a bedroom set in French style. It was sent to Paris for sale and was finally bought by a rich oil man in Oklahoma.

Some of the fine old mahogany furniture which graced the halls of ancestral homes years ago has been sold as antiques many times and in many cases has travelled here and there to all parts of the world. When some of the oldtime planters left their homes in the West Indies they gave their servants their big solid mahogany beds. I have seen a negro cabin more than half-filled with such a bed.

Mahogany is so well and favorably known that many other woods have stolen its name. It has been hard to convince people that the genuine, old-time mahogany is native to Florida; that it was in fact, common at one time here. Botanists have known this for many years. In the winter of 1942-1943 unusual cold weather killed unprotected vegetables, yet the mahogany was not injured, indicating that it might be grown even farther north than its present range.

The mahogany of Florida (*Swietenia mahogoni* Jacq.) is heavier than the mahogany of Mexico and Central America (*Swietenia macrophylla*), and the former is superior for the manufacture of slender parts of furniture, such as the legs of chairs. It has been used for many years for the rails of boats and of late in the manufacture of very solid revolving chairs for deep-sea fishing. The rich, reddish brown Floridian mahogany is often specked with small black spots which disappear to some extent as the wood darkens with age. For many people the beauty of the wood is enhanced by these spots.

The flowers of the mahogany are inconspicuous, yet they ultimately produce pods the size of turkeys' eggs. These open from the bottom, shedding many winged seeds, which look like the seeds of a maple tree. Of all our trees, few are greater seed-producers and few can germinate and grow on rough, rocky soil as freely as the native mahogany. The winged seeds, rotating as they fall, are so retarded in the descent that they are widely spread by the wind. Since cattle do not eat the bitter foliage of mahogany, it tends to gain ground in pastures.

The mahogany tree is not so slow to grow as is ordinarily supposed. Hard, heavy wood does not necessarily mean slow growth. The northern black locust is a hard and heavy wood and yet it is one of our quickest growers. I have known mahogany on mediocre soil in Florida close by the sea to add one inch in diameter each year. In rich moist soil it is a fast grower with a broad spread of dense foliage. When growing in an uncrowded location, it has massive limbs. These limbs are unexcelled for the ribs of boats as well as for furniture, because a beautiful crotch grain occurs where the limbs join the trunk.

Mahogany is a yardstick for other woods, and the tree in turn compares favorably with the live-oak for grace and sturdiness. After a recent cold snap some of the mahoganies on the mainland dropped their leaves. There soon followed a fine crop of pale green leaves, some tinged with red. This interruption in growth probably produced a distinct ring in the wood which would give it a variety of grain lacking in wood further south where growth is continuous.

A tea from the bark of the mahogany is used in fighting fevers. It seems to have precedence with the natives of the Florida Keys over all other bitter barks.

There are many little mahoganies hid by other trees in the jungle.

By cutting out the weed trees such as strangler-fig, poison wood and gumbo-limbo many of our key hammocks could be converted into mahogany forests.

There are thousands of acres of rough rocky limestone land close to the sea in Florida unfit for other things which can produce this king of furniture woods. It should be of interest to the state as a whole to put these acres to constructive work. It is the only place in the United States of America where this tree will grow, and a forest of mahogany is something which almost anybody would like to see. Land at present hardly worth more than ten dollars an acre ought to develop into something very valuable in the course of twenty years. Not only is it the best of timber trees but in a land of continuous growth you can get very tangible returns in a much shorter length of time. Almost every business requires time in which to get firmly established. Many have longer than twenty years in getting well under way.

From *Ten Trustworthy Tropical Trees,* pp. 35-39

Good trees must sow their own seed in abundance and then must fight fire, flood and drought.　　　　Ten Trustworthy Tropical Trees, p. 69

The Cajeput

In Australia this tree is known as the broad-leaved tea-tree. Its scientific name is *Melaleuca leucadendron*. The volatile oil yielded by its leaves and twigs is a valuable solvent and of great use in medicine. It is commonly applied externally in India for rheumatism. It is used, I have been told, as a basis for some massage creams, and as a remedy for toothache.

The bark of this tree is white and papery on the outside and corky and spongy to a depth of half an inch or more, even on small trees. It belongs to the same order, and is closely related to the genus Eucalyptus, but is superior to any Eucalyptus that I know of which would grow under similar conditions.

The seeds of this tree are unfortunately very minute and therefore difficult to sprout, in the usual way. (Scattering of these seeds over shallow water is all that is necessary. The minuteness of its seed is

therefore not a disadvantage since it can be started by the "flotation" method by shaking the seeds from a saltcellar from an aeroplane on the surface of any swampland covered with shallow-water). These seeds are, in fact, as fine as finely ground red-pepper.

Now close to a quarter of a century after introduction, this tree has gradually crept into its own and will probably in a very short time, with just a little help, change the landscape of the wet, mucky lands of South Florida. It also illustrates one of the great fundamental laws of nature that the living things which produce an enormous amount of seed are the ones most likely to survive and although millions of these seeds may never fall where they can germinate, enough will find a suitable lodging place to insure survival even against the destructive forces of nature and careless people.

It produces seed while very young. I have seen trees less than a year old bearing seed. The seed spreads on shallow water and when the water disappears germinates and grows quickly in the mud. The tree grows straight up and rarely assumes a bushy or shrubby form. Owing to the corky nature of its bark it is rarely killed by fire.

The long withes of this tree, somewhat similar to the olive, are excellent for garlands, especially because of the fragrance of its oily leaves. The tree is most excellent for decorative purposes for homes close to the seashore. Its tall straight white trunks are effectively used close to houses of certain types of architecture.

Wilcox in his book on Tropical Agriculture says that ten thousand pounds of cajeput oil is imported into this country every year. This does not signify much as it is often imported under other copyrighted names but, since it does enter into cosmetics, the amount consumed is probably much greater. There is always a demand for such oils for salves and nasal and throat sprays. I noticed that my dentist had a small bottle of cajeput oil and a small bottle of myrrh on his table, both of which he said he frequently used in his work.

A few pounds of this seed scattered from an aeroplane would do for a small outlay what would cost hundreds of dollars to do with less worthy species. In time it will do for the swamps of South Florida what the eucalyptus has done for Southern California and other parts of the world.

They say in Australia there is no malaria where it grows; if so there are probably no mosquitoes and if it repels mosquitoes by its odor or by drainage of the soil it would pay to plant it everywhere for that purpose alone.

I know of no introductions of cajeput in Florida previous to my introduction on the East Coast and the introduction by Andrews on the West Coast. Others may claim introductions but have no trees to show for them. The first plantings were on the bayfront at Coconut Grove and at Davie near Fort Lauderdale.

Back in the 1890's I was interested in the planting of Eucalyptus trees around the yellow fever hospital in Havana. It was then supposed, in several parts of the world, that emanations of volatile oils from the leaves offset the poisonous miasmas in the atmosphere. In fact malaria is an Italian word meaning "bad air." The trees probably sucked up the water from the limestone potholes and thus reduced the spread of malaria. After that the old Division of Forestry of the Department of Agriculture in Washington decided to try Eucalyptus trees on the edge of the Everglades. Several species were planted but none of them could withstand both flood and fire. This led me to write to Australia to ask if they had any other trees than Eucalyptus which were fast growing and at the same time both water and fire proof. In answer to this letter from the forestry service in Australia I received about a teaspoonful of cajeput seeds. With the help of Ed Simmonds, an enthusiastic tree specialist and a lovable character, I finally succeeded in sprouting these minute seeds.

Frank Stirling was interested in the planting of these trees at Davie. When they bloomed many people came to see the great masses of small white flowers. In time he established a cajeput nursery and usually has thousands of these trees on hand. He also noticed how the bees came in great quantities to collect the nectar. Then he specialized in cajeput honey.

Foresters like trees without inferiority complexes, trees that grow with vigor, giving rather than demanding protection. The land the forester uses is usually land unfit for other purposes. The less he has to spend the bigger will be his returns. He cannot spray and fertilize. He deals with the subsoil and with wild things. He is regulated by the law of diminishing returns. The sooner the land is covered and protected by a canopy the better. Good trees must sow their own seed in abundance and then must fight fire, flood and drought. They will at the same time add to the beauty of the landscape and the wealth of the nation.

The natural method of drainage is by the use of trees. Every tree is an efficient pump which never needs repairs or fuel for power. Trees consolidate the soil by their roots, add to the fertility by their decomposing leaves and furnish roosting places for birds that enrich the soil with guano. The best land in the Okeechobee region is the custard apple and elderberry bottoms. Nature has pointed the way and it is easy to follow.

Sudden artificial drainage changes the order of things. The equilibrium of nature is completely upset. Aquatic plants die and the fishes and other creatures, both plant and animal, that inhabit wet places are gradually replaced by other things. The birds that feed on these little plants and animals go elsewhere for food. Whether for better or worse, the landscape changes and new adjustments are in time established. It becomes an artificial arrangement under man's control, subject to all kinds of experiments and blunders, in his endeavors to improve on nature.

The tropics is primarily a tree country and not a country of herbaceous annuals, so there is really no marginal land except in treeless areas too wet or too dry for cultivation. The cheapest and best way to get this land out of the marginal realm is to plant it to trees of some kind, wherever possible. Reclamation by artificial drainage upsets the equilibrium of nature by lowering the watertable. It overdrains the surrounding highlands, and is, in short, a waste of time and effort unless urgently needed to supply food for an ever increasing population.

A friend tells me his cajeput blooms four or five times a year. There is therefore hardly a time when it is not producing seed. For that reason it is possible to collect the seed whenever needed. Break off the pliable fruiting twigs and put them in the sun in a quiet place on a newspaper and they will shell out of their own accord. Or better still put the twigs in a paper bag in the sun and the seeds will naturally come out. A little shaking of the sack now and then helps.

From *Ten Trustworthy Tropical Trees,* pp. 59-63, 67-71

Salt-Land Trees of the American Tropics

A halophite is a plant that grows in salt or brackish water. Halophytic trees are confined to the tropics where they grow by the seashore, or in salty places in near-desert regions. They are found on many thousands of square miles.

Halophytic trees are important because of the large areas they cover and also because many places close to the sea are growing saltier due to the digging of drainage canals. These canals lower the water table and the result is that salt water seeps into the land so that in time only trees that can endure salt will survive. The sea itself, as a matter of fact, gradually gets saltier, and in some places where I have been the tide rises and falls in wells that once were fresh. As the land gets saltier because of overdrainage, many of our plants will probably succumb.

Questions often asked about halophytes are these: Do they "prefer" salt? Do they feed on something contained in salt water? Do they, as a consequence of long-time adaptation, now require salt? Long ago, no doubt, the sea was lower than now. As it slowly rose, the plants which could endure salt water survived.

Certain trees *endure* certain conditions while others *demand* them. In forestry we have shade-endurers and sun-demanders. In the tropics we have sun-endurers and shade-demanders. Some of the halophytes grow in both salt and fresh water. As the land gets saltier because of overdrainage, many of our common plants fail.

Some of the halophytes, curiously, seem to "resent" the salt. These are similar in some ways to xerophytes, or drought plants. The red mangrove, for example, has thick, corky tissue around its roots, leathery leaves that reduce evaporation, and other characteristics of drought-resistant plants. Others, however, apparently absorb salt freely. The black mangrove (*Avicennia nitida*) has so much salt on the underside of its leaves that some people swish twigs of it in their cooking utensils to flavor their food. In plants whose roots are covered by salt water there is much osmotic action, as also in plants whose leaves are covered. You can easily determine the amount of salt that has been absorbed by chewing the tender parts of the plant.

Millions of acres of saline land are useless except for growing trees. Since they are free from fire and usually close to water transportation, they should be carefully tended and used by conservators despite the mosquitoes, sandflies, and crocodiles. Much of this land lies between the high and low water-mark and is classed as riparian rightage. When filled to a higher level, it becomes real estate. Places where you can travel in a boat, where oysters cling to roots of trees and mangrove snappers thrive, may seem like water to you, but they can be called forests since trees grow there with roots deep in the mud, and birds and orchids nestle in the branches. These forests, incidentally, are good protection for boatmen in severe storms.

Another and more important service is their constant construction and consolidation of muddy shores. The inhabitants of Dutch Guiana, remembering the methods used in the Old Country, have dyked some of the saline land to reclaim it for agriculture. Halophytes are grown there as pioneers for more tender plants. Sometimes the function of building land is carried on at the cost of the tree's own life. The red mangrove really kills itself; when, by the accumulation of detritus, the land is elevated a few inches, the red mangrove gives way to other plants. Along the shores of rivers it stops growing where the brackish water stops. Thus you can measure the salinity of the water by the size of the mangrove growths along the shore.

Before we discuss a number of halophytes, a word about their grains. Trees do not produce annual rings unless there is a distinct annual periodicity of some kind. Most halophytes grow continuously, and their grains are therefore probably the registration of spring and neap-tides. Perhaps, the full and dark of the moon are also written in the diary of the tree, since in wood everything that happens to it or around it is recorded by nature.

The following list catalogs some of the trees which are halophytic or near-halophytic:

Date-palm. The date is usually associated with the desert. It is, however, the tree of the oasis. It grows in salty soil close to bitter, brackish water and, as the Arabs say, "must have its feet in water and its head in the fires of heaven."

Coccolobis uvifera, sea-grape. This interesting tree belongs to the buckwheat family. It is a tough tree with stiff, coarse leaves and bears its fruit in grape-like bunches. Its upper leaves and twigs are often clipped off by the sand blast on the seashore. The tree is often toppled over but goes on growing. Its trunk and limbs are usually crooked, but the reddish wood is hard, tough, and heavy. Although the flesh on its seeds is scant, it is famous for jelly. It is considered highly ornamental by many people and ranks high as a protector against wind and moving sand close to the sea.

Conocarpus erecta, buttonwood. This tree grows close to the sea in muddy tidal marshes. It is our best fuel wood, yielding an excellent grade of charcoal. Both the bark and the wood produce the very best grade of tanning material. The wood is heavy, hard, and dark yellow-brown. It grows rapidly on salt muddy shores.

Laguncularia racemosa, white mangrove. This tree resembles the buttonwood. It belongs to the same family, grows in close assocation with it, and the wood has about the same characteristics and uses.

Bucida buceras, bucida, or black olive. This tree belongs to the same family as the last two mentioned. It grows in salt marshland, has hard, heavy wood like the white mangrove, and is also useful for tanning. It is a graceful, luxuriant tree and has been extensively planted in south Florida as a yard and street tree.

Rhizophora mangle, red mangrove. The red mangrove is probably the most pronounced of all halophytic trees. A greenish radicle drops to the water from the mother tree. It floats to some mudflat or oyster bar. The lower part forms a root; the upper end develops leaves. Other radicles do the same. Soon there are little islands. When the trees are a few feet high they send down branches which reach the bottom and there they root. More and more, these branches reach out, slowly invading the sea. Floating debris is caught in the labyrinth roots. The mud is consolidated and detritus accumulates. Little islands merge and soon a swamp results. In time, through the accumulation of debris, the level rises a few inches above the sea. Other plants get started from seeds floating in, some no doubt from seeds brought long distances in the mud on the feet of sea birds. Somehow in time other plants get started and the mangrove quits.

Avicennia nitida, black mangrove. Among the trees growing in the tropics by the seashore, I know of none more curious than the black mangrove. It was named in honor of a famous Oriental doctor, Avicenna of Bokara. It grows in great masses on the coast of South America, where it is known as *courida.* Its roots form great raft-like mats in the water. This is like a great mattress carefully constructed by engineers to control floods. It floats and moves up and down with the motion of the water. This mat sends up many brushlike structures a foot or two in the air. Like the knees of cypress, these may be for aëration. The seed sprouts at once when it hits the ground. The bark is rich in tannic acid, and the wood is heavy and hard. Although the black mangrove tree belongs to the same family as teak, the wood is coarse-grained and not very highly prized by the natives who use it.

The red and black mangroves are two of the strangest trees on the surface of this earth. They both grow in salt water and fight the sea. They are compatriots of tropical shores, and I can think of nothing stranger than to hear the ocean waves breaking into foam among the tangle of trees and see great masses of oysters clinging to their roots. Once in a tropical hurricane I sought refuge for my boat and myself in a cove among the mangroves. The boat was fastened by four ropes to the trunks of the trees. I could look up and see the clouds rush by. I could hear the water splashing through the forest. I was, however, safer there than at home.

Chrysobalanus icaco, cocoplum. This is one of the few trees of the tropics belonging to the plum family. It seems to be native to both the American and African tropics. This is a small tree which grows in dense clumps by the seashore and in both fresh and salt swamps. This tree is quite common around the edges of islands on the outer fringes of the Everglades. The Everglades proper is a vast bowl of muck covered with sawgrass, but on the edges, where it mingles with high land, many little hammock islands are found. These islands are usually fringed with a zone of cocoplum. This was an important fruit for the Seminole Indians. It is highly ornamental, yielding an excellent jelly from its plum-like fruits. Inside the seed, which is big compared with the size of the fruit, is an oily substance. The Indians say that this chocolate-like substance is very rich and nourishing, giving rise, no doubt, to the Bahaman name "pork-fat apple." I know of no plant more worthy of cultivation. The largest and best fruits of this species, according to several travellers, may be found in the Cayman Islands south of Cuba.

The generic name Chrysobalanus means "golden acorn," and the specific name *icaco* is probably old Indian.

From *Living by the Land,* pp. 75-83

Let us not fall into the common error of advising newcomers to a region to eat what the natives eat. Living by the Land. p. 90

Some people are more allergic than others to the juices of several trees that grow in the American tropics. The saps of two tropical trees— *manchineel* and *metopium*—are dangerously irritating to almost everybody. The juices of the *sand-box, cashew-nut* and *mango* usually cause only a mild skin irritation and fewer people suffer from their effects than from the effects of manchineel and metopium. The juices of all these trees, however, can seriously affect fair, tender-skinned people.

Irritating oils which are in these trees are carried in the smoke when the leaves are burned. Sometimes the wood and the hulls of the fruit retain these oils a long time.

Although many of the stories relating to the effects of manchineel and metopium sound almost fantastic, I know from my own experience over many years that some of the tales are far from fiction. I have known of many very painful illnesses and even of death resulting from contacts with manchineel and metopium. Despite my precautions, I have found it dangerous to collect branches of these two harmful trees for displays in my classes. (It is best to seal the samples tightly in cellophane envelopes.) Manchineel and metopium are common on beaches, and scantily clad people seeking shelter from the rain under these trees have developed illnesses which lasted for weeks; one person I knew barely recovered after being poisoned this way. It is interesting to note that American Indians used the manchineel to poison their arrows. Perhaps there is good reason for the old saying that he who sleeps in the shade of manchineel sleeps forever.

One of my neighbors who had made tea from metopium bark, mistaking it for gumbo limbo, was sick a long time, even though he discovered his mistake when he put the tea to his lips and washed his mouth immediately. Animals are said to be able to eat the fruits without danger; nevertheless, I have heard of people being badly poisoned after eating land crabs or fish which fed on metopium berries that had fallen into the water from trees growing along the shore.

Let us not fall into the common error of advising newcomers to a region to eat what the natives and the animals eat. Some natives eat what would kill a stranger to the area, and monkeys eat seeds that are poisonous to people. I remember seeing a mocking bird eat ten red hot peppers and then sing about it.

The poisons of these trees, as we have seen, affect people differently and the remedies do, too; what works for one person may be no good for others. Of course, there are numerous preparations with patented names for sale in drug stores, and almost every physician has his favorite prescription. People who are poisoned by these trees, however, are usually far from drugstores or doctors and are not interested in commercial concoctions. Once in a remote section of Cuba I was standing by a roadside when I merely touched the branch of a tree that I somehow failed to notice was a guao. First the palm of my hand turned black; then painful blisters appeared. Since I was far from any kind of store I had to search about for some antidote: I had Epsom salts and tea leaves among my supplies, and not far from where I stood I found an aloe leaf. I used all three freely with considerable success. The salts absorbed the water in the blebs, the tea leaves containing tannic acid probably coagulated the poisonous albumens, and the aloe–long famous as a dressing for burns–had a soothing effect. On another occasion I used some white marl, which is mostly a carbonate of lime, well mixed with glycerin to the consistency of putty.

Manchineel (Hippomane mancinella)

The name manchineel means little apple. In the old days when sailors from schooners, deprived of fresh food for many days at a time, saw fragrant fruits that looked like crab apples, they tasted them. Some lived and some died – and no doubt in this way the world learns what is safe and what is dangerous. But to this day newcomers are still experimenting and every now and then you read in the newspapers of the deadly effects of machineel.

On the Florida Keys and in the Bahamas it is only a small tree. In other parts of tropical America it reaches large size and yields a beautiful cabinet wood. To cut it down the cautious workmen pile trash around its base and kill it with fire. After it is dead and dry, the tree is cut and used – but, even so, only hard-skinned natives will work with it. Nature did so well in protecting the tree that, in consequence, it spreads where it pleases without the help or hindrance of man. In some places along seashores it has undisputed possession of the land. I think that the best way to get rid of it is to uproot it with a bulldozer and to keep going over the land with tractors until the manchineel disappears altogether. The flame-thrower used by our armed forces in wartime may have a peacetime use: It might be employed to advantage in ridding the land of manchineel along valuable seashore.

The name *Hippomane* was conferred on this genus by the great eighteenth-century Swedish botanist, Linnaeus. The word means *horse-rage* and was apparently applied to some plant in Greece used to excite horses. I have never learned what Linnaeus had in mind when he applied the word to our manchineel, unless it produced a frenzy in men similar to the horse-rage of years ago.

Metopium toxiferum and Guao

The metopium of the American tropics is not as poisonous as manchineel, but runs a close second. I have seen it growing by the wayside, in vacant lots, and even in school yards, in south Florida. It yields a big crop of blackberries and I presume that the birds which eat the seeds help the tree spread. If anything injures the bark of the metopium, the sap oozes out and turns the bark black. Popular names for the metopium are poison-wood or doctor-gum; in Cuba it is known as *guao de costa.*

From *Living by the Land*, pp. 89-92

9 Useful Plants for Florida

The lesson clearly is that we should stick to time-tested things. Things that are here, are here because they are fit.

Living by the Land, pp. 103-104

The Banana and the Pawpaw

I never cease to marvel at the banana and the pawpaw. Statisticians who have predicted a famine from the increase of population without a corresponding increase in the production of breadstuffs have neglected one potent factor—the banana.

The pawpaw or papaya is another succulent, quick-growing, prolific tropical fruit-producer, belonging in the same class of marvels with the banana, but is not related to it.

The banana is marvelous because of its prolific nature, yet it forms no seeds, and the great bunch of foodstuff when not used by man or other animals simply rots, and the stalk which produced it dies to give space to another to repeat the performance.

With me the banana is a favorite crop. I dig a deep hole in moist soil or muck. Into this hole I empty my waste basket containing old letters, newspapers, returned manuscripts, etc.; also the kitchen barrel containing tin cans and other stuff that the chickens will not eat; then

I throw in sweepings, rakings, old fertilizer bags, old iron, useless wood, bottles, and trash of any kind and every kind. On top of this I put a good forkful of stable manure and then some sand or muck. Then the banana root, often no bigger than your two fists, dry and lifeless-looking, after having been kicked about in the sun for a few days, waiting for planting time, is stuck into the ground and covered with a few inches of dirt.

In three months, if the weather is good, you may sit in the grateful shade of this big green-leaved plant. I almost called it a tree, because its stalk is as big as a man's leg and its foliage may be several feet above your head, but according to the definitions a tree must have a central *woody* axis, and to the banana there is no woody texture; it is all as soft as a cabbage and is usually completely consumed in a short time when left to chickens. Within a year a bunch of fruit is produced which a man can hardly carry—a bunch so big that it oftens bends the plant to the ground unless propped by forked sticks.

The most marvelous kind of banana culture may be seen in the Bahamas, on the Island of Eleuthera. Here there are deep holes called "banana holes" some of which are fifty or sixty or more feet in depth. At the bottom of these holes is moist rich earth. They are just like deep dry wells. A banana root is planted in a basket of soil, which is lowered with a rope to the bottom. The root sprouts and the stem shoots up like magic till it reaches the top of the hole. Then the foliage spreads out in the sunshine like flowers in a vase. There it grows and forms its bunch protected from the wind in the cool moist recesses of the hole. The bunch is formed at the surface of the ground, so that the enterprising native has but to pull it over with boat or sponge hook, sever it from the stalk with his machete, and walk proudly home with a week's provender for himself and family on his head—a fitting illustration of man's mastery over nature.

Little wonder that the native of the tropics is a lover of leisure; little wonder that he rests content in his palm-thatched hut amid his beloved bananas.

A good pawpaw [papaya] will bear a hundred or more melon-like fruits, a fruit to the axil of each leaf, ripe at the bottom and in all stages of development up to the bloom. The staminate and pistillate flowers are usually on separate plants, and the fruit varies a great deal in quality.

The fruit contains a large quantity of black, peppery seeds which may be removed *en masse,* as in the case of the cantaloupe. A good pawpaw, cold and treated with sugar and lime-juice, is relished by many people on a par with a muskmelon. The seeds are usually scattered in the midst of rubbish during the rainy season. As soon as the plants begin to bloom, all but one or two staminate plants are destroyed. In the course of a few months one may begin to pick pawpaws every day or so.

Of course some people have to learn to like them, but one lady that I know, of good habits, will steal this fruit when buying and begging fail.

Next in wonder to the prolific nature of this fruit is the marvelous fact that it contains a natural food-digester, a ferment now famous the world over as a medicine. Under various patent names it enters into the lists of many drug firms. By means of it men have already accumulated fortunes—not the producer, but the manufacturer and peddler who invent appealing names and have them patented.

I have before me a sample bottle containing one hundred pills for twenty-five cents. It is marked "Physician's sample. Our own preparation of the digestive juice of *Carica papaya* with willow charcoal." It is also marked a sure cure for dyspepsia or indigestion. I have often wondered where all this juice comes from. I have traveled in many parts of the tropics, but have never seen or heard of anybody collecting it, and the plant will not grow north of the frost line.

How fortunate the dweller in the tropics! If his meat is tough he can wrap it in pawpaw leaves overnight and it will be tender in the morning. If his meal has disagreed with him, he can step into his backyard and pick and eat a pawpaw for dessert.

Both bananas and pawpaws, however, are picked when full, but still green. This must be done to save them from the rats and birds. The tropical planter has bananas to roast and bananas to fry, sweet bananas and acid bananas, big bananas and little bananas, yellow bananas and red bananas—in fact, varieties galore.

The banana has been in a way the emancipator of the tropics. In many instances it has led the native out of thraldom. In many places from which bananas are not shipped he must work in the fields at a small recompense. At banana ports he can usually receive a cash payment for every full bunch. With bananas to eat and bananas to sell, the copper-colored native can rest in his home-made hammock, thump his home-made guitar, and smoke his home-made cigar with only one worry, and that is that he might at any time be forced to serve in the army of either the *de facto* or *de jure* government, for the cause of liberty. Even so he knows that the folks at home can live on the bananas and pawpaws and other fruits and vegetables growing in a semi-wild state around his bungalow.

From *The Everglades and Southern Florida*, pp. 38-41

Koonti

Koonti, or comptie, is a cycad of the genus *Zamia* with four species in Florida, according to Dr. John K. Small, a leading authority. The kind common in the southern end of the state is *Zamia floridana.* "Starch is very abundant in the underground portions of the plant, and it is often used for food," Dr. Charles Joseph Chamberlain writes in his book *The Living Cycads.* "The stem is pounded to a pulp and washed in a straining-cloth to remove a poison which is found in most cycads. During the Civil War several soldiers died from eating the root before the poison had been washed out; but the meal, when properly prepared, makes a fairly palatable cake or pudding.

But the plant was nevertheless important to the old settler. Whenever he needed cash he would manufacture a barrel of koonti starch, which was shipped by way of Key West and eventually converted into arrow-root biscuits. Pine trees were blazed for koonti tasks and it was common to see Negroes, Indians, and Whites all busy together, digging koonti. Here and there koonti mills were set up in the pine woods. The red water from the washings is poisonous to domestic animals, and the refuse, when it begins to decay, contaminates the air with a repulsive odor, although it is valuable as a fertilizer.

Just as a pioneer in another section of the region would depend on palm-cabbage and alligators' tails for food, the pine woods settler fed, without ill effects, on koonti starch and gopher turtles while clearing his land. As with several things in common use today, some people were probably poisoned before a safe process of manufacture was developed; however, this knowledge was absorbed by the early white settlers. When we plant any common cultivated thing we are profiting by the work of many people through many years.

The seeds of the koonti plant resemble large grains of corn and are called "koonti-corn." When eaten by turkeys or other animals they are believed to cause death, although some observers say that crows are not affected by the poison in the seeds. The leaves, resembling miniature palm fronds, were at one time shipped north to serve as decorations on Palm Sunday. Specimens of the cones of this plant are studied in northern laboratories because the pollen grains develop spermatazoa which wiggle about at a lively rate.

I once found what I thought were bacterioidal nodules in some piles of koonti roots. The presence of these nodules may account for the richness in nitrogen of the refuse and probably is one of the reasons why the plant never grows in low or damp places. Koonti is never seriously injured by forest fires, for it quickly sprouts from the roots if the leaves are killed. The only insect that seems to bother it is the koonti-fly, which seems to be immune to the poison in its leaves. Koonti is one of the left-overs of ages past, living only in small patches here and there throughout the world.

Koonti starch was the basis of sofki-pot, which was eventually superseded by corn grits. Grits replaced koonti, just as corn liquor replaced the black-drink or Florida maté, which the Indians called *yaupon.*

From *Living by the Land,* pp. 96-98

Florida Maté

Yaupon, the emetic holly, is described in D. J. Browne's *Trees of America,* published in 1857. In northeastern America it is referred to as *cassene, cassena,* and *cassioberry bush.* Its scientific name is *Ilex vomitoria.*

Ilex vomitoria is an excellent evergreen tree or shrub. In its natural habitat—moist, shady places from Virginia to Florida—it usually grows to a height of twelve or fifteen feet, but when cultivated it may rise still higher. The flowers which it puts forth in June are whitish. In October smooth red berries ripen, and they, like those of the common holly, remain on the tree's branches through the winter. A native of this part of the world, the emetic holly has long been known in England, to which it was introduced six years before our Declaration of Independence was signed.

It is said that *yaupon* was regarded by many of the southern tribes of American Indians as a holy plant. At any rate, they used it in their religious rites and solemn councils to clear the stomach—and the head. Each spring a chief would order the inhabitants of the town to assemble in the public house, but not before the structure had been purified by fire. After the tribesmen had convened, the chief was ceremoniously served a portion of yaupon broth in a new conchshell bowl. Then the other men were handed their emetic broth, and finally the women and the children received theirs. The beverage was supposed to strengthen the stomach and restore lost appetite, and men who partook of the ceremonial serving believed that in battle their agility and their courage would be increased. But in some tribes yaupon was held in such high regard that the decoction of its toasted leaves (called the *black-drink*) was thought too valuable to waste on women.

Early travellers in Florida tell in their memoirs about the eagerness with which Indians in the southern part of the state awaited the arrival of shipments of yaupon leaves from the northern sections. One legend has it that the name of the great Indian chief Osceola can be traced to the Seminoles' fondness for the plant. According to this tradition, the word Osceola is derived from *asi-yahola, asi* being the plant and *yahola* the long drawn-out cry that was heard when all the tribesmen were ready to drink. Osceola was often the leader in the drinking ceremonies; hence the name. Whatever the origin of his name, Osceola is remembered in our history books as the Seminole chief whose warriors fought against the United States Army in the 1830's.

For some time past, the Federal Department of Agriculture has been experimenting with yaupon in both South Carolina and Florida. It has been used as a beverage by white inhabitants of the Carolinas and Florida for many years. During the Civil War, when Oriental tea could not easily be imported, yaupon tea was used in the South. Today, inhabitants of the seaside swamps in North Carolina disguise the unpleasant taste of their drinking water by boiling in it a little yaupon. They use it constantly, as the Chinese do their tea. Like Paraguay tea, to which it is quite similar, it must be boiled, not steeped. It is more than likely that in many parts of the world, without these various teas, irrespective of their exhilarating effects, many persons would have died from drinking tainted water. Everywhere in

the tropics unboiled water is suspect. It is always best to rely on teas
and fruit juices.

From *Living by the Land,* pp. 98-99

Seminole Pumpkin (*Cucurbita moschata* or cushaw squash)

The Indians raised the pumpkin by planting it at the butt of a tree
which had been deadened. The vines climbed up the tree and into the
branches. It was a curious sight to see an old oak tree laden with
hanging pumpkins. The Indians may have had several reasons in mind
when they planted their pumpkins that way. The fruit was out of the
way of pigs and cattle; ground space was saved—the vine grew well
vertically; and the fruit was of a better quality than if it had lain on
the dank earth.

From the fact that this pumpkin is very much different from the
ordinary pumpkin, it appears to me that it has been cultivated for a
long time in one place, away from the chance of crossing with other
kinds of pumpkins. The Indians seemed to have selected the type best
suited to their needs. So far as I know, this pumpkin has never been
found in a wild state or even running wild in places where it is cultivated. I have seen pumpkins very much like this growing in the Bahamas and I suspect that the species was brought to Florida from the
West Indies long ago. But the method of growing them on trees is, I
think, a Seminole invention.

This pumpkin is small, hard, and greenish, and in general it does
not look very promising. Nevertheless, it really has the best flavored
flesh of any pumpkin I have ever tasted.

About forty years ago I visited an island in the southern part of the
Everglades. The Indians there did not appear to be Seminoles. They
fitted descriptions I had read of the aboriginal Calusas, and these may
well have been a remnant of the old Calusas who gradually merged
with the Seminoles from the North. The center of the island I visited
was covered with live oaks which had been girdled. Large numbers of
pumpkins hung from the branches. The Indians also had a kind of
dasheen; they grew cassava in the fertile ground under the oaks, and
close by they had a patch of bananas. This led me to believe that
somehow they had been in communication with the West Indies, for

south Florida at the time was well separated from the rest of the state by miles of mud and by unbridged rivers.

Unlike the majority of Seminoles I then knew, these Indians were rather unfriendly and very difficult to approach. Finally the members of my party won over the women by giving them some big black cigars. After that we were allowed to look around the settlement, and by the time we were ready to leave the Indians thought well enough of us to give us each a pumpkin. I planted the seeds in my garden and raised fine pumpkins for several years, even though I did not grow them on trees. Everybody who tasted these pumpkins in pies or other forms remarked about the fine flavor.

The Seminoles favored the pumpkin not only because it is tasty, but also because it can be cut into strips, dried, and eaten in times of dearth. For good reason the pumpkin was a very important part of the Indian diet; the Seminole chief Billy Bowlegs was mad when surveyors trampled his bananas, but he was fighting mad when they shot down his pumpkins. Without a doubt the Seminole pumpkin is so far superior to the common run of pumpkins that it could be cultivated and sold or canned with profit.

The pumpkin was the Seminoles' favorite fruit, always hanging from the tree limbs, protected by its hard shell and its height above the ground. Often it is necessary to cut the pumpkin open with an axe and shell out the inside as we do the coconut.

From *Living by the Land,* pp. 100-101

Florida Quinine

Various forms of malaria constitute our worst tropical diseases. As a means of controlling malaria and other fevers, bitter barks have been used in the tropics for many years. For this purpose plants belonging to the madder family have long been famous.

The efficiency of quinine in the treatment of malaria is generally accepted. Manufacturers insist, of course, that no substitute for quinine can be successfully used; but, in time past, when unscrupulous dealers adulterated quinine, daring physicians used—and with considerable effectiveness—a crude, powdered Peruvian bark in place of the standard preparation.

In wartime, quinine has always been scarce. During the Civil War importation of products from abroad was seriously curtailed, especially in the South. During that war my old friend Dr. Charles Mohr, a druggist in Mobile, searched ceaselessly for suitable native substitutes for the unobtainable foreign drugs. Instead of quinine he used the bark of the Georgia fever tree, *Pinckneya pubens.**

Elderly Southerners may recall, from stories they heard as children, that at regular intervals slaves on the old plantations were lined up and required to take their dose of fever-bark. Perhaps some of the slaves did not look upon this health measure as an ordeal—the fever-bark was soaked in rum.

Now very scarce, the tree was once plentiful in the swamplands. It has white showy flowers tinted with red. Professors Coker and Totten, in their excellent book on the trees of the southeastern United States,** declare, "Pinckneya is a close relative of the cinchona tree of South America that furnishes the quinine of commerce and probably contains the same curative element, as its effectiveness in curing malaria has been repeatedly proved."

Personal experience has made me certain that not only quinine but several other bitter barks are excellent preventatives of malarias of various sorts. The amoebae that cause the fevers do not flourish in the body of a person saturated with these bitter drugs. Some day someone will discover that Georgia fever-bark (*pinckneya*) is quite as good as South American cinchona, perhaps even better.

Another tree belonging to the madder family is princewood (*Exostema caribaeum*). It grows on the Florida Keys, where the natives fight fevers by drinking water in which princewood bark has been soaked. Soaked in rum, the bark's medicinal properties would be even more effective. It might be possible to manufacture a salable bitters from this tree. In fact, both the Georgia fever-bark and the princewood should be tested as preventative medicines without delay, since the time is not far off when they will both be exceedingly rare. Already they are scarce and in time they will completely disappear from Florida unless protected.

Indians have lived in the Western Hemisphere a very long time and

[*Also known as Florida quinine *(Pinckneya pubens)*. Ed.]
  **William Chamber Coker and H. R. Totten, *Trees of the Southeastern States* (1937).

learned a lot from experience. It is not a mere coincidence that the words *experience* and *experiment* come from the same root, *experiri*, meaning to try out. The Indians unquestionably tried out many things through the ages, the results of which they passed on willingly, unwillingly, or even unknowingly, to us newcomers.

From *Living by the Land,* pp. 104-106

Billy Bowlegs, Chief of the Seminoles. From a photograph by Clark of New Orleans. Frontispiece in Billy Bowlegs and the Seminole War.

10 Seminole Indians

The best way to understand anything is to study it from its beginnings.   Rehabilitation of the Floridan Keys, p. 58

About 1860 the Seminole abandoned buckskin pants and coats for highly colored calico kilts, better suited for wading in the glades. The brave Chief Osceola's paternal grandfather was a Scotchman, and his mother married a Scotchman. This, together with the comfort of such light and inexpensive dress, may have caused the change. They also abandoned log huts for lighter and healthier living quarters. They neglected their cattle when they no longer had Negro slaves to do the chores.

The word Seminole is Creek for "runaway," but they have another long name for themselves meaning "people at the point of the land." As the whites needed the land, these Indians were forced southward into the glades. They absorbed other Indians, some white blood, and some Negro blood in the process of this migration.

According to Roy Nash,* who has recently studied these Indians

[*Roy Nash was an agent of the United States Indian Service. Dr. Gifford accompanied Mr. Nash on a government inspection tour of the Seminole reservation and filed a report with the Bureau of Indian Affairs, Department of the Interior, Washington, D.C. Ed.]

for the Commissioner of Indian Affairs, they [the tribe] earn about fifty thousand dollars per year mainly from alligator, raccoon, and other skins. This income could be easily increased, but about thirty-five thousand dollars of this money goes to bootleggers. The Indian can get his meals out of the swamp and has left about one hundred fifty dollars per family per year for tobacco and gunpower. The Seminole therefore is not only self-supportive but helps support several white men. The exploitation of the Indian has always been a fruitful occupation.

These Indians could be gradually induced to move to or near the Everglades National Park to serve as guides and guards. They could be formed into a sort of constabulary to watch over the Park. They would be very useful in tracking lost persons or even runaways in hiding.

We can never repair the damage done the Seminole. Why begrudge him a home and hunting ground in such a lonely place, where no white man except a fugitive from justice would dare to live?

It is a long story, but they put a price on his head, then in a friendly way gave him rum, and when drunk captured him, and against his will shipped him to Oklahoma. Worse that that, there seems to be good evidence to prove that to get possession of his homeland, early settlers disguised themselves as Indians, raided their own settlements to stir up strife and draw in soldiers from the north.

Even the officers in our army praised his honesty and bravery. Leave him as he is. Let him roam the glades at will. He is a part of the picture of the Florida of long ago.

It is safe to say that very few white men have ever fathomed the mind of an Indian. No psychologist can get his intelligence quotient. He is a harmonious element in the landscape. He never dominates it as does the European. He lives in harmony with all created things, a philosophy that the white man could follow with profit.

Away from the open glades and prairies of Florida, of which he seems a vital part, he acts lost, in fact, is lost, and will in time be overwhelmed by the preponderance of the whites around him. Although there are no mountains in Florida, there are mountains of clouds. Instead of desert sands, there are green glades dotted with glistening lakes and watercourses teeming with fish and water fowl. Here and there are islands of cypress or pine, but the picture will lose a vital part if the Seminole with his little family in his dugout canoe quits it forever.

From a broadcast over radio station WIOD, April 16, 1931.

I know the location of the grave of an Indian in one of our college towns. He came from the Far West to get the education of the white man. But it was too much for him. He died of loneliness and home-sickness. Nostalgia or homesickness is a terrible disease. It has killed as many soldiers as bullets. It is most severe in those who have been reared in an isolated place with a close circle of relatives and friends. It is not a desire for home altogether, but for the environment of which one is a part. Fear, dread, ignorance and suspicion of the white man entered into it. Although the hut he lived in could be built in a day, the Indian had it, as is strongly evidenced by the acts and words of both Tiger Tail and Coacoochee. The thing that makes nostalgia bitter is compulsion, and this was applied without stint to the Indian.

From *Billy Bowlegs and the Seminole War,* p. 35

Billy Bowlegs' Grandson

Recently I had the pleasure of spending a couple of days with the present Billy Bowlegs. He says he is sixty years old and is the grandson of the Billy Bowlegs of Seminole War fame. Although not very talka-tive he is far less taciturn than the common run of Indians. He has the reputation of being a pretty good Indian. He goes on long journeys afoot throughout the southern part of this state, walking twenty-five or thirty miles a day. He was very much amused when I called him a "tourist." He told me that Okeechobee City was his post office. He had his teeth filled with gold. He took most excellent care of his dog. His dog was a real Indian dog and apparently understood no English and had the same quiet sleuthful and reticent character of the Indian. He bought meat for his dog and prepared for him a fine bed of Florida moss under our automobile. He said his two horses were dead but that he still had two wagons. He owned a canoe, a gun, a blanket and some household utensils. A very few dollars would buy all his worldly pos-sessions. His dog, canoe and gun practically constituted his estate. His dog was evidently his best friend and constant companion and his family if he had any gave him very little concern. He said he had no squaws at present and from the way he said it, I inferred he was glad of it. He had a single-barrel shot gun of inexpensive make which he prized very highly. He had no desire for any superfluous clothing; he ate very little and with raccoon hides at more than two dollars each he could live from day to day with no worries and slight privation. By eating sparingly, by walking long distances in the open air, and by keeping away from the contagions of cities he will probably live to a ripe old age and finally curl up and die under the palmettoes on some little Everglade island and if found be buried with some of his belongings without any doctor's or undertaker's bills to pay or any bother over wills or inheritance taxes.

Although he seemed very much civilized in some respects; I am sure he would not steal and if he promised to meet you at a certain point at a certain time he would be there. He said he raised a few vegetables now and then but hunting was his business. Lots of white men have gone into this business in Florida and talk more of their dogs and guns than they do of their families. Just as a boy in the process of growth passes through the various stages of culture through which the race has passed so does the white man under proper conditions quickly revert in the same way. The wildlife of the glades which persisted in great abundance during Indian times quickly vanishes when the white man takes hold. The Indian never kills just for the fun of it. He kills for food or for money with which to buy food. The pot-hunter has good reason for killing but the man who kills for fun is in a class by himself even if he is white.

The whole of South Florida should be one big bird sanctuary. All hunting should be stopped. Tropical birds would come here in great quantitites and bird lovers would come here to live from all parts of the world. Hunting gives pleasure to only a small portion of the population. Wild bird life like beautiful plant life gives pleasure to every normal person.

Billy Bowlegs when not acting as a guide collects otter, raccoon and alligator skins for the market. I think he resembles the picture of his grandfather and is on the whole an improvement on his distinguished ancestor.

Chief Tigertail

I have not seen much of the Chief Tiger Tail in print. The story is told that rather than be deported he pounded glass into powder and drank it with water and died in consequence.

The following is Canova's account of the death of Tiger Tail: "Amongst our captives was Tiger Tail. He was determined not to leave Florida alive, and committed suicide while at Myers in a horrible manner. The morning they were to leave, he procured a quantity of glass which he pounded fine, and swallowed in a glass of water. While the guards were taking him down to the dock where the steamer lay waiting, he told Sampson the negro interpreter, that he was going to die. At the same time he asked the guards to let him lie down, which they permitted him to do. Spreading his pallet upon the ground, he laid himself on it and in a few minutes, with the Indian's stoical indifference to the pain he suffered, and to the approach of death, he died. His daughter was with him, and when he breathed his last she threw herself upon his dead body, wailing so piteously that the by-standers, men used as they were to death and sorrow, could not keep back their tears.

Tiger Tail was buried at Fort Myers in the land he loved better than his own life. Whatever his faults were he was a brave chief, and valiant-ly defended the land in which he was born, and which he felt was his birthright, and it was fitting that he should find his last resting place in its bosom, where all, red men and white are the same when they sleep the sleep that knows no waking in this world."

A tall monument should be reared in memory of the man who died rather than leave the state of Florida. When a man is so fond of his native land that he would rather be dead than live elsewhere is an advertisement for the land of palms that would be hard to rival. The name Tiger Tail is still common with the Seminoles and one Icha Tiger Tail, the little daughter of Jack Tiger Tail who was murdered on the Miami River not long ago, died recently in a Miami hospital and was buried in the City Cemetery.

Although there are hardly more than five hundred Indians in South Florida there was a time no doubt when it supported a much larger population. There are evidences on all sides. Kitchen middens contain-ing many potsherds and other Indian remains are common. These pots were brought from elsewhere since there is no clay in the southern part of this state fit for pottery. Here and there are big mounds and other evidences of these people of the past. Here and there are piles of shells on the shore where they feasted year after year on oysters, conchs and other shell fish. In one place I know of there is a little round, apparently artificial harbor, surrounded on all sides by piles of shells. It is easy to picture these canoes in a circle on the beach and the Indians on the crest eating oysters much larger than those of today.

The Green Corn Dance

Let me add a few words in reference to one of their most important if not the most important of all Seminole ceremonies—the Green Corn Dance. It was of course much more than a dance. It was a solemn annual religious festival of the Creek Indians but similar to many of the festivals that most peoples of the world celebrate. They purge themselves with various herbs, wash themselves, start new fires, rid themselves of sin and filth, repent and start afresh with all kinds of little ceremonies each with a potent meaning. It seems to be a sort of combination of New Years, Salt-water-day, Spring-house-cleaning, Camp-meeting, etc. They seem to be trying to rid themselves of sin by doing all sorts of stunts except whipping themselves as do the Flagellantes of New Mexico. Bartram says when a town celebrates the busk (from Creek puskita, a fast), having previously provided themselves with new clothes, new pots, pans, and other household utensils and furniture, they collect all their worn-out clothes and other despicable things, sweep and cleanse their houses, squares, and the whole town of their filth, which with all the remaining grain and other old provisions, they cast together into one common heap and consume it with fire. After having taken medicine, and fasted for three days, all the fire in the town is extinguished. During this fast they abstain from the gratification of every appetite and passion whatever. A general amnesty is proclaimed, all malefactors may return to their town, and they are absolved from their crimes, which are now forgotten, and they are restored to favor.

According to this there is fasting and pardoning of criminals and forgiveness of sins followed by feasting and rejoicing over the fruits of the year because the new year begins when the crops mature in late summer. They start with new pots and new clothes and new fire. It is really a remarkable institution because it combines in one at least half a dozen ceremonies, religious and otherwise, that the civilized peoples of the world indulge in today. I have never seen this dance but have heard and read much about it and am sure the significance of much that happens during these days of fasting and purging and feasting is known only to the Indian.

From *Billy Bowlegs and the Seminole War*, pp. 51-55, 77-78, 175

11 Chips and Splinters

Northerners and Habits

To succeed in living in the tropics, settlers from the North must adapt themselves to new surroundings rather than attempt to change the surroundings to suit themselves. I have seen northerners in the tropics dig cellars under their homes for cool storage and put steep roofs on their buildings, as if they were needed to shed ice and snow. Attics are also useless in the tropics, except for storage space. Tropical natives sometimes make the same error of trying northern ways in southern climates. Once in the West Indies I saw a high school student diligently reading a government bulletin which told how to preserve fruits for winter use. Many tropical boys and girls, whose textbooks originate in the North, are perplexed by descriptions of spring and summer wood. In many places in the tropics you can find northerners who are trying to impose superior northern ways in their new homes. Hence they plant groves as they do in California or central Florida. They are showing the natives how they do things back home.

Although the native of the tropics may seem backward in many ways to the outsider, the native knows from long traditions of his people how to live in harmony with the land. More than a century ago

Noah Webster wrote, "Tradition occasionally hands down the practical arts with more precision and fidelity than they can be transmitted by books." Webster's statement is unquestionably true about agriculture, woodcraft, and the domestic arts. On the other hand, tradition may work undesirably. Many years ago in Europe I noticed that the natives in one section were irrigating their fields by one method, whereas a few miles away farmers working with the same topographical and climatic conditions were using another method. In following lessons their fathers had taught them, they probably never thought of comparing results with one another.

Land, plants, and people must work together, wherever they are; else disaster comes soon or late. The simple, unsophisticated native may be untutored in book-learning but he can show the new settler a thing or two about living by the land.

From *Living by the Land,* pp. 20-21

On Planting a Garden

It is often said in Florida that the man who smokes a pipe, is good to his dog, goes fishing and plants a garden never goes to Raiford or Chattahoochee. The former is a prison, the latter an insane asylum. Although it is seldom lit, the pipe smoker is usually meditative. If good to his dog, he is good to other things. Fishing develops the habit of watchful, patient waiting; but if he plants a garden of trees he reaps a hundredfold in more ways than one, helps the section in which he lives and does something very worthwhile to himself which is hard to describe. There is no easier or cheaper production than when a man with the help of his family, raises what he needs for his own use on his own land.

From *Ten Trustworthy Tropical Trees*, p. 10

Tropical Yards

The lawn is a northern institution.　　　Living By The Land, p. 110

From my point of view green lawns do not belong in the tropics. Northern grass, accustomed to lying dormant in winter, cannot be used in a climate where continuous growth takes place. Again, many places in the tropics are too wet, others too dry, for ordinary grass, and anything delicate enough to give the desired lawn effect is usually crowded out by hardier plants. Among the low-growing legumes we might be able to find a covering for the ground better than any grass, a kind that will keep out other plants and at the same time protect and enrich the soil.

In Florida, grasses are cut and raked and presently the soil is impoverished. Expensive fertilizer and fresh top soil are applied, but in time chinch bugs, army worms, and other pests come. (We must remember that unless grass is well kept it is worse than none.) The property-owner who starts out to have a beautiful lawn finds, instead, that he has a first-rate headache. Lawns in the North are easily maintained for centuries; in many tropical countries, on the other hand, the almost insurmountable difficulties have led many men to abandon the idea of having a verdant lawn.

Clearly, as I have noted, there is nothing really tropical in a green lawn. The attractiveness of the tropics, I have always thought, lies mainly in the brilliant colors. We have plenty of green in the tropics, in forms other than grass, but we never have enough flowers or fruit anywhere. Tropical climate demands bright colors.

Residents of tropical Hispanic-America who understand the area in which they live do not try to maintain lawns. As a matter of fact, they ignore also the northern tradition of placing a house in the center of a yard—their houses surround the yards. The patios are pleasant parts of the home, with their flowers, paths, and properly arranged shrubs. If the patio is large this is enough in the way of a yard, and the land outside the house is left to go its own way or it can be planted with crops of the region. In the southern sections of the United States, a landscape architect with vision might design, for a home in which no patio can be built, a yard that would have flowers, shrubs, and paths, but no lawn. Perhaps a fenced-in garden of flowers and low shrubbery could be arranged around the house; outside the fence let the field or the forest hold sway.* If we can ever get away from the northern lawn

---

*Of course it is necessary to keep the weed cut, although, as the world-famous horticulturist, Lincoln Bailey, used to say to his classes at Cornell University, "A rosebush is a weed in a wheat-field and wheat would be a weed in a rose garden." The water hyacinth is one of the most beautiful of flowering plants and at the same time a pernicious weed. It broke loose long ago and got into our rivers and canals, where it is matted so thickly that the water is no longer navigable.

tradition we might be able to develop something better by studying the gardens of tropical peoples. Lawns embellished with iron deer or concrete watchdogs ought not to be hard to improve on.

My advice that the northern lawn tradition be abandoned in tropical sections of the United States is not directed at people who can afford the cost of top soil, fertilizer, water, and manpower in fighting against natural conditions. I am speaking to the man of moderate means who tries to follow suitable traditions and maintain a normal standard of living. Few people are courageous individualists like a few of my acquaintances. One man I knew cemented his yard and painted it green. A woman I know pulled up the grass in her yard and swept the ground carefully now and then with a brush broom. Another friend of mine cut down all the trees on his property because grass would not grow in their shade and because the falling leaves littered the lawn. Well, a home should reflect the personality of its owner. He is the one who should have it the way he wants it, remembering that safety, comfort, and beauty should be combined as much as possible.

Never forget that the lawn is a northern institution. When a tropical country is settled by northern people they should change to fit their new environment, not try to change the environment to fit them.

From *Living by the Land*, pp. 107-110

The Use of Sink-Holes

A hole in the ground is an asset rather than a liability. You should look upon it as a valuable aid in growing plants, not as an unsightly or even dangerous nuisance that must be filled with dirt at considerable expense. Admittedly, in areas subject to frequent low temperatures a hole in the land is nearly worthless for purposes of cultivation because it is appreciably more vulnerable to frost than the surrounding land. But in the tropics you have no such problem. Here, the deeper you dig, the warmer and more moist the earth becomes, and the greater the protection from wind and sun. In some places these holes are called lime-sinks or simply sinks; in Spanish-speaking countries they are known as *cenotes*; in the Bahamas they are termed banana holes.

Organic matter naturally accumulates in these depressions. The acids produced when the matter rots dissolve the carbonate of lime down to the level of the normal water table. These natural compost reservoirs draw warm water from the bottom, if they are deep enough. During heavy rains they quickly absorb excess water. In rocky lime-stone they are excellent drainage agents.

The value of a hole in lime-rock country seems to me so great that I recommend the creation of holes wherever natural ones do not already exist. If you do not have a natural depression in your back-yard, blast a hole as big as your land permits. The resulting loose rock can be used in the building of houses, walls, roads, and paths. The rocks can be piled around the bases of trees with good results, for the dampness under them will benefit the roots. If, however, a large amount of the blasted rock is soft and you have no other use for it, mix it with trash and throw it back into the hole. The hole should be filled with rakings and organic matter of any kind to form a natural compost heap. In the warm, moist hole disintegration progresses rapidly and continuously throughout the year. As the substance set-tles more organic matter should be added.

Bananas, papayas, pineapples, and other fruit plants thrive in sink-holes. The rich decayed matter in the hole nourishes the plants, which yield a surprising amount of fruit.

The many advantages of having a truly sunken garden are apparent: you gain cheap, readily available rocks that can be used in construc-tion; you have a place where you can dispose of any kind of debris, including old cans, scrap iron, and even old newspapers; you utilize otherwise useless trash; above all, you will get remarkably good yields from your fruit-bearing plants growing in the compost. Since plants set out in sink-holes grow in thick patches, abandoned rock quarries and natural sinks which are thoughtlessly spoken of as liabilities can be converted into ornamental and productive features of the home landscape.

If there is no natural sink-hole on your land, make one! It pays.

From *Living by the Land*, pp. 73-74

Unripe Fruit

We should never tolerate the picking and shipping of unripe, im-mature fruit. The juice of unripe fruit may be actually injurious to health. And certainly all the work that goes into the development of fine fruit is wasted if we pick it when it is still green.

Nature herself cautions us against taking fruit before it is ripe. The seed lies in the heart of the fruit. It contains the embryo of the plant, carefully wrapped and packed for travel. Inside the seed, food material is stored to serve the plant until it can get its own. The seeds have many devices which effect their spread. On the outside of many seeds, once they are ready for distribution, are sweet pulp, palatable juice, fragrant aromas, and bright colors attracting the various animals that will eat the fruits and then cast the seeds where they can take root. Before the seeds are ready for distribution, on the other hand, nature repels those who would start the seeds germinating. Unripe fruits are an inconspicuous green, dry instead of juicy, acrid or even poisonous— altogether unattractive to hungry animals. Not until the proper time does the fruit advertise itself. Then it tells the world in unmistakable language.

The desire to avoid unripe fruit and to have tree-ripened fruit throughout the year will bring new residents to the South. In time the planting of homesites with diversified food trees for home use (not for sale or profit) will become popular. The shipping of green fruit will force many people to grow their own, to dry or can their surplus, or to convert it into food pastes. I do not think that I overstate my case by saying that sheer disgust and self-defense will plunge many people into growing their own supplies of fruit. Obviously, a defense against high prices would be the production of one's own tree crops. Nowhere else in this country can we have foods in more constant variety, freshness, and abundance than in the South.

From *Living by the Land,* pp. 48-49

Twigs and Sticks

In America we must get away from our worship of size. Small sticks of wood are just as much as tremendous logs. The fact that they are little does not change their character or their chemical composition. Twigs and sticks contain cellulose, just like sapwood, which is used in pulp; in fact, cellulose belongs in the same chemical category as other sugars. Alcohol is a product of sugar fermentation; formalde-hyde is the product of wood alcohol and oxygen; phenol is also a product of wood alcohol distillation. Grind up wood and add hydro-chloric acid and you get wood-sugar-syrup. By drying, filtering, centri-fuging, you get sugar; and by fermenting you get alcohol from the sugar. Glycerine is produced from vegetal fats—the most important are copra and other palm oils—and animal fats of various kinds; *but all these other things are, or can be, produced from wood cellulose.*

Wood fibers can be used industrially in making paper, rayon and celanese, celluloid, collodion, guncotton, and lacquer.

The great lesson of conservation, I think, is this: *nothing is valuable if we fail to use it properly.*

The measure of value is not only knowledge but also the proper *application* of knowledge. "Know-how" is more than just knowing; it includes doing. To get this "know-how" in utilizing nature, we must have the right kind of schools for the purpose. I have long advocated that every university establish a school of conservation.

From *Living by the Land,* p. 116

Plants that Eat Animals

Animals are in general supported by plants. A few plants, however, feed in part at least upon animals. A tree in New Zealand called the bird-catching tree, *Pisonia brunoniana,* apparently has no use for the birds it catches. Perhaps in time they will learn to leave this tree alone. There is a species of Pisonia on our Keys. It is a vine that is very catching in its ways. It has the descriptive names of "devil's claws" or "pull-and-hold-back-vine." It is armed with vicious recurved thorns and is very detrimental to anybody who is in a hurry. The specific name, *aculeata* means provided with prickles. This genus belongs to the common garden four o'clock family, which includes also the South American bougainvilleas so common throughout the tropics, which are also well-armed with stiff sharp thorns.

Some time ago an expedition was sent from England to Madagascar to study a strange plant. The natives would cast a sacrificial maiden into the heart of this tree. Its slimy leaves would close upon its victim, and death and disintegration soon followed. These sacrifices of maidens are in my mind questionable. It is hard to imagine any people no matter how superstitious or religiously zealous who would be willing to get rid of a good looking woman that way unless she had it coming to her. This idea of throwing a maiden into the middle of a carniverous tree is beyond belief. So far there is no record of any carnivorous tree anywhere on the face of this earth.

In the Far East there is a tree called the Upas which has a poisonous juice. The very word "upas" means anything exceedingly deadly. According to early tales the exhalations from this tree were such that both animals and plants in its vicinity fall dead, that birds flying over it fell dead, and that a desert surrounded the tree. This tree has since been cultivated in hothouses and botanic gardens without ill effects to anybody. In fact, the natives use the bark to form sacks to carry rice. The idea of its fatal exhalations may be because it grew in Java where carbonic acid or perhaps carbon monoxide gas escaped from crevices in the earth which suffocated creatures that passed over it. The juice of the tree is, however, so poisonous that it is used by the natives to dip their arrows.

These stories appeal to the imagination to such an extent that they are hard to down. Nature is marvelous enough without embellishments. A few years ago a young scientist, just in fun, described a bug that endangered life by boring holes in the steel of railroad tracks out West. He explained how it worked from the underside and went unnoticed till the rail was broken by the weight of a passing train. This story was copied and recopied many times until it finally died a natural death.

Plants, of course, use both vegetable and animal organic matter in the form of humus after it is dead and in the process of decomposition. Except for putrefaction and consumption of this dead organic matter it would accumulate mountain high over the face of this earth. In places where decomposition is slow because of cold and other conditions not favorable to decomposition it does accumulate in great northern beds of peat and duff in which even the mastodon is preserved, even to its flesh and hair; the longest cold storage known to man. Except for the wind and the force of falling water on the face of the earth our great source of power comes from the vegetable world. It is the machine that renders the energy of the sun available to man and other animals.

There is a very curious group of plants that feed upon insects. They would digest flesh of any kind that falls their way. They can be fed small pieces of meat. This supply of nitrogen supplements what is garnered from the soil by the roots and from the air by leaves. I refer to a group generally classified as insectivorous plants. The fact that they capture living creatures is not so wonderful as the way they accomplish it.

The simplest are the sundews. They have stiff hairs on their leaves. On the tip of each hair is a sticky substance. When an insect gets tangled in this mucilagenous substance the hairs close in on it and even the leaf enfolds him so that he is soon digested and absorbed. In the case of the venus fly trap, the leaf closes and soon devours the fly. The bladderworts so common in water in the swamps by our roadsides capture small crustaceans in an underwater bladder device. Pitcher plants, Indian dippers, and other plants have leaves shaped like cups into which the animal falls and cannot escape because of stiff hairs which point downward. Some contain water in which the animal slowly drowns, while others have a roof over the top to protect it in heavy downpours, yet open enough to permit the entrance of little animals of various kinds. It almost seems that these animal-eating plants have a sense of feeling. Feeling is the most fundamental and less local of all our sense, the one which we could least afford to spare, and it may be that this is the one sense possessed by all living protoplasm in primitive form.

These animal-eating plants are not very abundant and not very large. It is a sort of a reversal of the ways of nature, a little experiment apparently well-planned but never very successful, and expensive because it naturally fails to fit into the general scheme of things.

The modest little sundew, nature's flypaper, would appear monstrous if enlarged on the screen while crushing and engulfing a mosquito in its slimy digestive fluids. The picture of a butcher bird impaling a fat wiggling worm on a thorn might be barred from showing because of its cruelty.

Man is about the only animal that kills just for the fun of it. This is not inherited from the lower animals or is not a survival of the animal instincts within him because animals with very few exceptions never kill more than is needed for their immediate use. To kill to live and not to live to kill is the order of the natural world. In that way a perfect equilibrium is established among living things which would go on undisturbed forever were it not for the great upheavals occasioned by man's interference. Each sacrifices a part so that all may live. Each finds its own niche in the order of things, otherwise it fails to survive. It is coordination and cooperation as much as fight and competition.

From an unpublished manuscript, 1935

Harney's Punchbowl

For many years in what is now called Brickell's Hammock there has been a famous round hole about the size of a sugar barrel in the rock, under an overhanging shelf of limestone, called Harney's Punchbowl.* It is protected from the sun and conveniently located on the shore of the bay. A cask could be easily filled and rolled to the shore.

This spring was named after General Harney, the famous Indian fighter. It dates way back to the earliest days. For all we know it might have been Ponce de León's Ponchera. Water was scarce on all the keys. The Indians to the south suffered at times for lack of it, at other times from too much of it. They squeezed the juice from cactus fruits into depressions in the rock and drank.

The pool was as necessary to the vagabonds that sailed the seas of this section as to the planters, and it is safe to assume that Harney's Punchbowl has had a very checkered history that has never been recorded. It is difficult to distinguish between pirate and planter. The peaceful backbreaking work of the planter often was only a camouflage to hide his real occupation. Like wild creatures that congregate around a waterhole to drink or patiently wait their turn, both planters and pirates peacefully partook of this refreshment. It was a cool, healthy waterhole for the common good.

It serves no purpose now. It is simply a relic which in time will be forgotten and lost, but here as everywhere throughout the world, especially in regions where water is precious, these waterholes marked the difference between life and death, success or failure.

When I first knew the spot there was an old house where an English clergyman by the name of Bolton sold jellies to visitors. The trees

[*Historic freshwater well, now dry, located along the shore of Biscayne Bay just south of the entrance to the Rickenbacker Causeway. Ed.]

were festooned with night-blooming cereus and native vanilla was common there. Large rattlesnakes hid in the rocks, and wildcats were common. A man still living in Miami was attacked by a young cougar that jumped on his back from the limb of a tree. Sea cows and alligators were common, and one old-timer, while swimming, was attacked by an alligator. He saved himself from mutilation and death by following the old Indian custom of sticking his thumbs into the creature's eyes.

Soldiers camped near there during the Spanish American War. They amused themselves by carving the overhanging rocks. Near the punchbowl there were some big strangler figs. Their roots like boa constrictors had enveloped other trees and were creeping like snakes through crevices in the rock. Here and there still may be found the broken pots of the Indians.

The Everglades water is gone. It no longer seeps through the limerock to the edge of the bay. With this water go countless creatures that depend on it. Somebody aptly called this hole in the rock a punchbowl, probably because of its shape, probably also because it was better than the best punch to the thirsty seaman.

From unidentified newspaper article in Dr. Gifford's Blue Scrapbook

Pot-Holers and Rock-Pilers

Let us begin by forgetting all that has been handed down to us by northern methods of soil cultivation in our treatment of many acres of limestone land in the American tropics. Tradition is strong and difficult and many no doubt say forget these barren islands and deal with lands which can be plowed and harrowed and scarified so that we can do as our fathers did. First of all these lands have been depleted in this very way and since they are well located in healthful islands where the white man can live and be happy and since they are capable of producing crops of fine quality, they are worthy of redemption. Many of these limestone islands are located in our front yard and were once richly covered with vegetation as is evidenced by root-cores, historical documents, and other sources of information. Bermuda and Barbados are examples of what can be done. They were once considered hopeless but now support a large population with much besides for export. When burnt over many times these limestone islands become covered with hard crusts of limestone. The heat converts the surface rock into quick-lime which slacks when it rains, finally hardening again into glassy rock. On Big Pine Key I have seen the natives breaking the rock surface into soil with a sledge hammer in order to plant a small patch of okra, peppers, and tomatoes.

After years of observation I have divided these lands and their cultivators into two groups: Pot-holers and Rock-pilers. There are two kinds of pot-holes; those that are produced artificially with dynamite, and those that are of natural formation. A hole is drilled in the rock to the depth of about a couple of feet. Dynamite is packed in and exploded and the rock is shattered sometimes to the width of a yard from the center. This rock is pulled out and used for wall or house construction or left in piles near-by. The holes are filled with sand or marl or whatever can be found in the way of soil. This is the artificial pot-hole. In the explosion some of the rock is converted into milk-like liquid which is forced into the pores around the sides of the hole. This soon hardens into rock again and destroys the porosity of the rock unless the hole is filled with disintegrating organic matter or the hole is gradually widened by other charges of dynamite. All this is expensive and in the meanwhile the little tree is subjected to a series of rough experiences. If you keep widening the hole with dynamite, and pulling out the rock, the holes come together in time and you have done nothing except to lower the level of the land in general.

In the case of natural rock holes, sometimes called lime-sinks, banana-holes, cenotes, or rockpits, the formation starts where the rock is softer or where there is a depression. Organic matter collects, forms acids which eat away the rock down to the normal level of the water table. If this is not brackish, it is inhabited by many kinds of living things and is very fertile. It is a fine place for such mammoth herbs as the banana. The big leaves of the banana are not frazzled by the wind, and as is well known, tattered leaves do not produce fine fruits. These are the places normally sought for and cultivated by the native pot-holer.

Now the reverse of this is the rock pile. Years ago I noticed that the mounds constructed of rock by the Indians were covered with a dense vegetation. These mounds were not for burial or ceremonial purposes. They were built as a place of refuge during hurricane overflows. The rocks were carefully placed with dirt in between to help bind them together. I have noticed in other places how Indian ruins consisting of loose rock were completely covered with dense jungle growth.

I have also noticed how with some trash mixed in with the rock, vegetation developed in rock piles. Seeds blown in by the wind or dropped by birds soon sprouted in the crevices of the rock. Coconuts in the pile sprouted and since they produced roots throughout their stems, gained a strong foothold in the rock pile. We all know that tree roots need air, but they need water also and a rock pile seemed to be dry. One night by a large rock pile I heard water dripping. It was dew and there is nothing more invigorating to plant life than dew. Dew never forms unless it has something to form on and those loose rocks afforded a large surface for its collection. On still, cloudless nights, close to the sea-shore the dew is very heavy. As old gardeners were wont to say—a good dew every night is better than a light shower in the daytime. Then I tested the temperature. When I put the thermometer in the open sun I cannot say how hot it got because the mercury went to the top of the glass but I do know that the temperature in the rock pile was many degrees lower than in the open.

Then I began to pile rocks high around my trees. They kept the roots cool, they collected dew, and they held the tree in place in times of storm, but they did more. In between these rocks countless creatures of many kinds found a home. They were many kinds of ants, cockroaches, spiders, centipedes, scorpions, rats and snakes. They, I found, left behind a very rich detritus which fed the tree. Anyway old lime trees which were dying gained new life and productivity. The limes were more abundant and richer in juice and now it is quite the custom to pile rocks around the trees instead of hauling them away or leaving them in piles between the rows to be in the way. Thus the rock pilers come into being. As the lime limbs bent over to the ground and touched the rock, I piled rock on the limbs. Where they were scarred they rooted by a process of layering, producing new young trees that bear profusely while very young. All people said what a tough, inhospitable tangle you have produced in a lime grove which is always difficult to approach because of its spines. Anyway in spite of difficulties, rock-pilers are forming. In northern climates a rock hole would be a frost hole, but in the tropics the deeper you go the hotter

it gets. I have always opposed anything spiney, but when they produce such fruits as the Key Lime, it is worth the hazard of rocks and
spines. It is usually so that the best things on earth are hard to get.
Anyway they are the things we crave. Maybe prohibition is the very
thing meant to stimulate desire. It seems strange that many trees have
spines to keep away the very things that spread their seeds. Many say
the spines are there to protect the fruit until it falls to the ground.

From "Pot-Holers and Rock-Pilers," *Organic Gardening,* 1946

Cracker Boys' Dinner; 'Gator, Cabbage Palm

Live oak acorns and cabbage palm heart nourished the Seminoles
for ages; in Florida's early days poor Cracker boys found cabbage
palm soup and alligator tail a good hard-times fare, also.

The live oak and cabbage palmetto are unquestionably the two
most distinctive all-Florida trees, and rightly so since by their sterling
qualities they command the adoration of everybody. Without the
moss-covered oak and groups of cabbage palmetto Florida wouldn't
be Florida.

There is the stately royal and the useful coco palm but there is a
homespun loveliness and peculiar environmental fitness in the cabbage
palmetto. It was and still is dear to the heart of the Indian and early
settler, because its terminal bud furnishes them with a delicious vegetable necessary to the health of all who lived mainly on fish and game.

The bears of Florida are also fond of it, and many trees have been
killed for a poy of cabbage. Modern man must have it canned. It may
be found on the menus of high-class restaurants at a high-class price. It
is nevertheless criminal to kill a tree for a mess of cabbage.

From unidentified newspaper article from Dr. Gifford's Blue Scrapbook

Conchs and Crackers

For many years the bulk of the all-year population of Florida
consisted of Conchs and Crackers. Outsiders partook of the climate,
but in the early days the Conchs and Crackers ruled the land. The
great out-of-doors was theirs. Little by little the so-called tourists have

settled. Little by little the old time Conch and Cracker have been outnumbered. Only here and there do they still hold sway and the time is not far distant when they will be described in history as the early settlers, as picturesque groups of vigorous and independent pioneers, who developed customs all their own, well fitted to the land of Florida as it once was. They were ever-willing to share the climate and the beaches with the Yankees in winter, but it was all theirs in the summertime.

The Conch clung to the seashore. He had no use for the backland. He was hospitable and courteous. He was also very religious, but if a wreck occurred on Sunday he excused himself for a time to be in it. He could sail a boat intuitively and if need be dive like a turtle. In the early days the spongers and fishers gathered in some snug harbor on Sunday. On one of the largest boats designated by the American flag at masthead they sang and prayed all day. He had houses of peculiar architecture, surrounded by limes and sapodillas, pineapples, coconuts and other fruits, close to the sea where he could watch for wrecks. Some had ports in their garrets and widow-walks. His main tools were the machete and bowie knife. He weeded with a knife. They say he was called Conch because he loved the shellfish. It was an important part of his diet, but so were other things that live in the sea. He was of English extraction and came to the Florida Keys by way of the Bahamas. He was an English creole. A creole is a person of European blood born in the American tropics. At one time in Nassau they had an uprising. In the absence of flags they stuck conch shells on poles as their insignia. They could file off the tip of the big end and produce a horn which if properly blown would roll out in melodious sound for miles over the ocean. Their conch-horn signals were like an unwritten code, but in some parts of North Carolina the poor white whites were still called Conchs. This would indicate that the word had another meaning. It is also more than likely that the word Conch dates way back and that its association with the shellfish is only accidental.

The conch was pulled out of its shell, tied in little bundles and hung on nails on the side of his house in the sun and air. They treated fish in the same way, and even relished cooked fish heads which are ordinarily saved for the cat. Although all these things were plentiful, close to their doorsteps, they were salted and kept for a few days. Conch stew or chowder is very rich, nourishing food. The shells were often cut into cameos. When burnt into lime they formed a binder equal to the best grade of cement.

The Conch settler has merged with the rest of us and practically is no more.

He was very different from the Cracker who lived in the backwoods up the state although both were of old English stock. One wrestled with the sea and the other fought the forest. The Cracker was as much a part of the Florida of yesterday as the cabbage palmetto or moss-draped live oak. In the early days in Virginia the first families stuck to the rich valleys and lowlands and owned slaves to do the work. The poor people scattered into the hills and on southward and for many years remained about the same. They were of pure Anglo-Saxon stock, set and stubborn in their ways, so that they changed very little in the course of time. They finally reached Florida and were to the pine forest of this state what the Conch was to the seashore.

Fifty years ago central and north Florida was almost a solid mass of virgin timber. The roads were sandy in places with bogs and mud in between. The ox was the main motive power. The turpentine industry moved south and the Cracker moved with it. Logs were carried or dragged by strings of oxen. These oxen were controlled by a long rawhide whip with a hickory handle. An expert teamster could crack his whip with a loud report so that it echoed throughout the woods. They soon learned to signal and then actually talked with one another with these whips. It was like the drums of Africa. My most vivid impression of the Southland as a young man was the resounding of these whips throughout the thick colonnades of massive pine trees. They were so expert they could brush a fly from the ear of an ox or kill a rattler or send a hound dog howling through the woods. They prided themselves on the quality of these rawhide whips. It was a weapon, a utensil and a means of communication. In groups they vied with one another in all kinds of whip stunts. And this no doubt is why they were called Crackers.

They were a forest people. They needed oxen for the naval stores and lumber industry so they developed cattle for the purpose, and some of these cattle with wide-spreading horns were probably direct descendants of the breed left behind by the Spaniards.

The Conchs had turtle and sponge crawls; the Cracker had his cattle corrals. They let their stock roam at will through the forest and prairies. They fenced in their own patches of collards, cane and corn. The great out-of-doors belonged to everybody. The cattle were branded and many of the old cows lived and died in the swamps where they were born. These customs became so fixed and the right to pasture their cattle and pigs anywhere developed into what they conscientiously thought was a right. These rights as populations increased became troublesome. In these modern times with wonderful highways and speeding autos, roaming cattle and pigs are a danger. The highway was not intended as a parking place for cattle and hogs. The pasture is always sweet by the wayside. For years the trains throughout Florida suddenly halted and their whistles shrieked. A glance from your window and you would see cattle with tails on high bounding into the swamps. They were a constant worry to engineers and a dead cow at a road crossing surrounded by buzzards was a common occurrence. People who have anything worthwhile to protect must build a fence that is bull-strong, hog-tight, and horse-high.

The Conch and Cracker lived on the natural resources of the great out-of-doors with little or no restraint. That is now about ended with the exhaustion of these resources. It may last in remote districts a few years longer. The old must pass before the new begins. The sons and daughters of these old settlers are a part of us and the younger generation knows full well that the days of pioneer privileges are past.

Many of the Conchs were descendants of loyalists from the United States who settled in the Bahamas. The Crackers were those who could not or would not leave. The two meet again in the state of Florida with changes in character due to their environments—the one a seafaring people, the other a people of the forest.

From unidentified newspaper article in Dr. Gifford's Blue Scrapbook

12 Remember the Dodo

The time to preserve anything is while it is still plentiful. Living by
the Land, p. 133

Almost everybody recognizes that the extinction of animals and
plants, unless they are clearly destructive, is a regrettable thing. Once
gone, they are gone forever.

The demand in the past for animals to be shown in zoos and
circuses has been great. The zoos are fine educationally for showing
city-folk the infinite variety of nature, but wild animals usually do not
flourish in captivity. Another force working toward extinction of
some species has been the big game hunters, men who are sometimes
genuinely interested in sport but who are more often eager for tro-
phies and material to write about. Pursued by exhibitors and marks-
men, and hunted down because their pelts are valuable or their meat is
good to eat, many kinds of animals are now almost extinct. In addition, men have changed the environments of wild creatures and thus
destroyed them unwittingly. We should protect all animals and plants
from extinction, at least long enough to discover their value to the
present and future generations.

Insects are not all destructive. Two species—the honey bee and silkworm—have been domesticated and are enormously useful to people. Even termites are not all bad. Although they will chew up a house in short order, they rid the forest floor of slash and rubbish. Of the many thousands of species of insects, only a small proportion are really destructive. The good kinds often feed on the bad, and the insects are food for many beneficial creatures. Every species of living thing is a part of the economy of nature; its presence or absence means better or worse for its fellows.

Clumps of trees and bushes are less bothered by insects than solitary trees are. Trees standing together furnish a good cover for birds, lizards, and toads, all enemies of insects. Also, trees growing close together are usually hardier than those standing alone.

Spraying trees with poison, unless carefully done, kills all bugs, good and bad. When you kill one kind of insect, another kind is likely to appear.

To make up for the destruction wrought by hawks and owls, many kinds of birds benefit men. Waterfowl that feed on fish are often of great value to us as depositors of guano. In general, however, the kinds of birds that most deserve to be protected are those that live in cleared fields or in thickets near the habitations of men. Wrens, quail, redbirds, robins, mockingbirds, and woodpeckers nest in barns, chimneys, and sometimes under porches. Many orange growers have found that a covey of quail will keep down scale in a grove. Most berry-bearing trees grow from seeds that have been distributed by birds.

Humming birds in the American tropics and sun birds in Africa carry pollen from flower to flower. They have long, slender bills that enter deep into the flowers for their nectar. Their plumage has developed a sparkling color rivaling the beauty of the flowers they visit. In the tropics, certain plants rub their pistils and stamens against the feathers of birds that rest upon their branches.

A bird that has been much maligned but deserves commendation and protection is the woodpecker. Besides, the woodpecker should have special notice here because no bird is more closely associated with trees, and no bird has comparable wood-working machinery.

The woodpecker grasps the bark of a tree and braces himself with his twelve expanded tail-feathers. Then he begins pounding with his head and chisel bill until he has excavated a home for himself. His head and neck are wonderfully constructed to endure the constant, strenuous drilling into wood.

When he searches for food his vigorous, regular pounding on dead trees echoes through the forest and terrorizes every insect in the region. When a woodpecker knocks on a metal pipe on the roof of your home, the bugs scamper just as chickens run for cover when they see an owl or hawk. A dead tree is usually a menagerie of living things. There the woodpecker feeds on insect larvae in the rotted wood. He searches for bugs in the crevices of the bark, and he tears off dead bark to get to the larvae of burrowing beetles.

The woodpecker's tongue seems designed to reach into deep holes. It is long and mucilaginous, and it spears its quarry on little fishhook barbs. The woodpecker rolls his tongue back into a cavity in his head when he is not using it. Few birds are so marvelously constructed for the work they must do to get their daily food.

One variety of woodpecker in our Spanish Southwest is called the *carpintero,* the carpenter (the Spanish common name for all woodpeckers). The carpenter woodpecker drills a hole in the bark of an oak and carefully places an acorn in it. After a time the acorn develops a grub. When that happens the woodpecker returns to the meal he has prepared for himself.

The woodpecker is very important in keeping woodboring insects in check.

Now the sapsucker, which belongs to the same family, is of no value to the forest. His search for food injures living trees. The sapsucker digs little holes in perfect circles around many trees. He feeds upon the sap and the insects that get caught in it. His tongue is like a paint brush, wiping the hole clean as he circles the trunk of the tree for his meal.

When the sapsucker penetrates the cambium layer of the trunk, adventitious buds are produced. A bud is a growing tip surrounded and protected by nascent leaves or scales. Normal buds are either terminal (at the end of a shoot) or axillary (in the axils of the leaves). Adventitious buds produce eminences and irregularities in the wood, which, if sawed in a certain way, may give a kind of bird's-eye grain.

Where sapsuckers have worked for a long time, the bole of the tree sometimes becomes enlarged from the constant irritation. Fungi enter the holes bored by the sapsuckers and rot commences. In time the tree dies. Some trees, however, suffer no apparent ill-effects from the sapsuckers' holes, but apple trees in the North have been killed by them.

Why should anyone want to kill a woodpecker? A sapsucker, perhaps—but not a woodpecker! The woodpecker is not good to eat. He
is a poor songster. A western variety has a disagreeable odor. But none
of these failures is a capital offense. We need woodpeckers to help us
save our trees.

Pileated woodpeckers always bring activity and noise. I am always
thrilled at the sight of a flock of flamingoes or roseate spoonbills or
sandhill cranes, but I find the greatest pleasure in seeing a group of
pileated woodpeckers.

At the top of the woodpecker hierarchy are the ivory bills. The
noted ornithologist Gilbert Pearson says that in addition to the protected colony of ivory bills in Louisiana, colonies of ivory bills probably can be found in southwestern Florida. Ivory bills have become so
scarce that positively no shooting of the birds and no taking of the
eggs should be allowed, even for scientific study.

The ivory bill is almost gone; the pileated woodpecker will follow.
When these two varieties are exterminated we shall have lost two of
the most interesting members of the bird world. Unless we do something soon to preserve them, we shall certainly lose a great many other
birds which were once abundant. The time to preserve anything is
while it is still plentiful.

From *Living by the Land*, pp. 130-133

Notes

1. F. Page Wilson, in a letter to Dr. Charlton W. Tebeau of the University of Miami on October 20, 1949, wrote regarding a memorial to Dr. Gifford: "In local and national comment on his passing, a good deal has been said about the use of trees and conservation, tropical forestry, etc., but not much about the broader aspects of his teaching. That is, on the many deep-seated but subtle differences of this Bay Biscayne-Keys-Caribbean country and their relationship with human living. It seems to me that, were we to seek for Dr. Gifford a short-phrase, inclusive name, it well might be, 'This region's Great Interpreter.' "
2. John C. Gifford, "*The Natural Resources of Florida, I.* Geographical Location," *Everglade Magazine,* circa 1913. Clipping from Dr. Gifford's Blue Scrapbook.
3. John C. Gifford, *Living by the Land,* p. 20.
4. Dr. Gifford classified southern Florida's climate as tropical-wet-dry, identical with patches of the West Indies, Central and South America, a micro-tropics on the American mainland distinct from any other climatological region in this country. Another prime indicator of the tropics used by Dr. Gifford was the presence of the coconut palm and other native and introduced tropical species. Among his contemporary naturalists who claimed South Florida as tropical, David Fairchild also considered the coconut palm his prime indicator. Thomas Barbour noted the presence of tropical spiders. Charles Torrey Simpson defined as truly tropical any area where a "majority of the native plants have been derived from the Torrid Zone." A. H. Howell, ornithologist, cited the many tropical birds common to southern Florida.
5. Wadislaw Gorczynski, world climatologist, in *Economic Geography,* places

this region with the "best" climates that cover only 1% of the continental areas of the world. See Bibliography under "Carson."

6. John Kunkel Small (1869-1938) was a naturalist associated with the New York Botanical Gardens, 1909-1938. See Bibliography.

7. Thomas Barbour (1884-1946) was a naturalist, former head of the Museum of Comparative Zoology and the Peabody Museum, Cambridge, Massachusetts; and a Director, Fairchild Tropical Garden. See Bibliography.

8. Gifford, *Living by the Land,* p. 133.

9. Gifford, *Living by the Land,* p. 58.

10. Gifford, *Living by the Land,* p. 15.

11. David Fairchild (1869-1954) was a noted plant explorer, head of the section of Foreign Seed and Plant Introduction, U.S. Department of Agriculture, and a resident of Coconut Grove for many years. See Bibliography.

H. L. Nehrling (1853-1929) was a noted Florida naturalist whose work was mainly carried out on the west coast of Florida.

Charles Torrey Simpson (1846-1932), conchologist and naturalist associated with the United States National Museum in Washington, lived in the Little River section of Miami. See Bibliography.

12. Bill Mabry was a carpenter and resident of Coral Gables, Florida.

13. Lewis Adams, a lifelong friend of Dr. Gifford, established two of the largest book manufacturing firms, including Kingsport (Tennessee) Press and the Colonial Press (Clinton, Massachusetts). A millionaire, he had an estate on Biscayne Bay and spent much of his time there until he died in 1968.

14. John C. Gifford, *The Everglades and Southern Florida,* p. 31.

15. Gifford, *Living by the Land,* p. 133.

16. John C. Gifford, *Rehabilitation of the Floridan Keys,* p. 58.

17. *Forests and Forestry in the American States,* Ralph R. Widner, ed., compiled by the National Association of State Foresters, 1969.

18. John C. Gifford, *Floridan Keys with Special Reference to Soil Productivity,* p. 52.

19. George Merrick was the developer of Coral Gables, Florida, one of the first planned communities in America, opened in 1921.

20. *Forests and Forestry in the American States,* Ralph R. Widner, ed., compiled by the National Association of State Foresters, 1969.

21. John C. Gifford, "Recollections of a Faculty Member of the First College of Forestry," *Society of American Foresters Proceedings* 2(1948):333-335.

22. The first official organ of the American Forestry Association.

23. Gifford, *Floridan Keys,* p. 52.

24. Louis Agassiz (1807-1873), a noted Swiss-American zoologist and geologist, was a Harvard University professor whose ideas influenced a generation of American scientists.

25. Gifford, "The Luquillo Forest Reserve, Porto Rico."

26. Gifford, "Recollections of a Faculty Member."

27. Gifford, "Recollections of a Faculty Member."

28. Gifford, "Recollections of a Faculty Member."

29. Gifford, unidentified newspaper article, no date. From Dr. Gifford's Blue Scrapbook.

30. Ralph Middleton Munroe, yachtsman, designer, and ship builder, settled in Coconut Grove in 1881. See Bibliography.

31. Commodore Munroe and others record that the Seminole Indians were frequent and welcome visitors to their homes.

32. Gifford notebook.
33. John C. Gifford, *Ten Trustworthy Trees,* pp. 67-68.
34. Lewis Adams' term.
35. John C. Gifford, *Tropical Subsistance Homestead.*
36. Unidentified newspaper article from Dr. Gifford's Blue Scrapbook.
37. Gifford, *The Everglades and Southern Florida,* p. 85.
38. Gifford, *The Everglades and Southern Florida,* p. 84.
39. John C. Gifford, *Reclamation of the Everglades with Trees,* p. 28.
40. Gifford, Talk to the Homestead (Florida) Home Beautification Group, May 1933. Clipping in Blue Scrapbook.
41. Unidentified newspaper article, Blue Scrapbook.
42. Gifford, "Looking Ahead," *The Tropic Magazine,* July 1914.
43. Nineteenth governor of Florida, 1905-1909.
44. Gifford, *The Everglades and Southern Florida,* Preface.
45. Ralph M. Munroe, *The Commodore's Story,* pp. 241-242.
46. *A Guide to Miami and Dade City.* American Guide Series. 1941.
47. See *River of Grass* by Marjory Stoneman Douglas for an account of the 1926 hurricane.
48. Henry Troetschel, Jr., "John Clayton Gifford, An Appreciation," *Tequesta* 10 (1950): 35-47.
49. From Robe B. Carson's eulogy to Dr. Gifford in the *Miami Daily News,* Sunday, July 24, 1949.
50. Gifford, *Tropical Subsistence Homestead,* p. 15.
51. Gifford, *Living by the Land,* p. 24.
52. Gifford, *Rehabilitation of the Floridan Keys,* p. 43.
53. Gifford, *Rehabilitation of the Floridan Keys,* p. 66.
54. Gifford, *Tropical Subsistence Homestead,* p. 16.
55. Gifford, *The Everglades and Southern Florida,* Chapter 1.
56. The Fairchild Tropical Garden is the major tropical botanical garden and tropical plant research center on the United States mainland. It is located just a few miles from Dr. Gifford's former home in Coconut Grove.
57. Gifford, Unidentified newspaper article reporting one of Dr. Gifford's speeches. Blue Scrapbook. No date.
58. John C. Gifford, *Floridan Keys,* p. 28.
59. Gifford, *Rehabilitation of the Floridan Keys,* p. 24.
60. John C. Gifford, "Farm Forestry," *Proceedings of the Florida State Horticultural Society,* 1936, p. 115.
61. Gifford, *Living by the Land,* p. 24.

Bibliography

American Tree Association (Washington, D.C.), comp. and ed. *The Forestry Almanac.* Philadelphia: J. B. Lippincott Co., 1924, 1926.

Blackman, E. V. *Miami and Dade County, Florida. Its Settlement, Progress and Achievement.* Washington, D.C.: V. Rainbolt, 1921.

Ballinger, Kenneth. *Miami Millions. The Dance of the Dollars in the Great Florida Land Boom of 1925.* Miami: Franklin Press, Inc., 1936.

Barbour, George M. *Florida for Tourists, Invalids and Settlers.* New York: D. Appleton & Co., 1883.

Barbour, Thomas. *That Vanishing Eden.* Boston: Little, Brown & Co., 1944.

Brookfield, Charles M., and Griswold, Oliver. *They All Called It Tropical.* Coconut Grove, Fla.: Data Press, 1964.

Canova, Andrew P. *Life and Adventures in South Florida.* Tampa, Fla., 1904.

Carson, Robe B. "The Florida Tropics." *Economic Geography* 27 (No. 4, Oct., 1951):321-339.

Cohen, Isidor. *Historical Sketches and Sidelights of Miami, Florida.* Cambridge, 1905.

Dorn, Mabel. *Under the Coconuts in Florida.* Miami: South Florida Publications, 1946.

Dorn, Mabel. *Tropical Gardening for South Florida.* South Miami: The South Florida Publishing Co., 1952.

Douglas, Marjory Stoneman. *The Everglades, River of Grass.* New York: Rinehart & Co., 1947.

Dovell, Junius E. *The History of Banking in Florida.* 1955.

Early, Donald M. "Tropical Florida Viewed through Naturalists' Eyes." Master's thesis, University of Miami, 1964. (Southern Florida as viewed through the writings of a half dozen noted naturalists who lived and worked in this area from 1881 to 1954.)

Fairchild, David G. *Exploring for Plants.* New York: Macmillan Co., 1930.

Fairchild, David G. "Florida's Climate and Soil." *Fairchild Tropical Garden Bulletin* 8 (No. 3, 1952).

Fairchild, David G. "Our Plant Immigrants." *Geographic* 17 (1906):179-201.

Fairchild, David G. "Reasons for a Large General Plant Introduction Garden in Southern Florida." *Proceedings of the Florida State Horticultural Society* XLVIII (1934):117-120.

Fairchild, David G. "Reminiscences of Early Plant Introduction Work in South Florida." *Proceedings of the Florida State Horticultural Society* LI (1938):11-33.

Fairchild, David G. *The World Grows Round My Door.* New York: Charles Scribner's Sons, 1947.

Fairchild, David G. *The World Was My Garden.* New York: Charles Scribner's Sons, 1938.

Florida East Coast Railroad and Hotels. An Illustrated Guide. 1903.

Fontaneda, Hernando d'Escalante. *Memoir of D^O. d'Escalante Fontaneda Respecting Florida.* Spain, ca. 1474. Tr. by Buckingham Smith, Washington, D.C., 1854. Reprinted with revisions, David O. True, ed. Miami: Univ. of Miami and Historical Assn. of Southern Florida, 1944.

Forests and Forestry in the American States. Ralph R. Widner, ed. Compiled by the National Association of State Foresters. 1969.

Forestry Almanac. Compiled and edited by the American Tree Association. Philadelphia: J. B. Lippincott Co., 1924, 1926.

Gifford, John C. For a listing of his writings, see pp. 205-216.

Gifford, John C. and Rodale, Jerome I. *Stone Mulching in the Garden.* Emmaus, Pennsylvania: Rodale Press, 1949.

Hollingsworth, Tracy. *History of Dade County,* Florida. Coral Gables, Fla: Glade House, 1949.

Hosmer, Ralph Sheldon. "Forestry at Cornell." Pamphlet published by Cornell University, Ithaca, New York, December, 1950.

Hudson, F. M. "Beginnings in Dade County." *Tequesta* 1 (No. 3, 1943).

Lummus, J. N. *The Miracle of Miami Beach.* Miami: The Miami Post Publishing Co., 1940, 1944.

Menninger, Edwin A. *Fantastic Trees.* New York: Viking Press, 1967.

Morton, Julia. "The Gifford Arboretum." University of Miami. *Bulletin No. 3, Gifford Society of Tropical Botany* (1952-1953):10.

Morton, Julia. *Wild Plants for Survival in South Florida.* 2nd ed., rev. Miami: Hurricane House Publishers, Inc., 1968.

Muir, Helen. *Miami, U.S.A.* 2nd ed. Miami: Hurricane House Publishers, Inc., 1963.

Munroe, Ralph M. and Gilpin, Vincent. Reprint from 1930 edition. *The Commodore's Story.* Miami: Historical Assn. of Southern Florida, 1966.

Nehrling, Henry. *My Garden in Florida.* Estero, Fla.: The American Eagle, 1944.

Nehrling, Henry. *The Plant World in Florida.* Collected and edited by Alfred and Elizabeth Kay. New York: Macmillan Co., 1933.

Parker, Alfred Browning. *You and Architecture.* New York: Delacourt, 1965.

Parker, Martha. "Biography of Dr. John C. Gifford." University of Miami. *Bulletin No. 1, Gifford Society of Tropical Botany* (1950-1951):3.

Redford, Polly. *Billion Dollar Sandbar: A Biography of Miami Beach.* New York: Dutton, 1970.

Reese, J. H. *Florida Flashlights.* Miami: The Hefty Press, 1917. (Miscellaneous facts about Florida.)

Simpson, Charles Torrey. *Florida Wild Life.* New York: Macmillan Co., 1932.

Simpson, Charles Torrey. *In Lower Florida Wilds.* New York: G. P. Putnam, 1920.

Simpson, Charles Torrey. *Observations on Flora and Fauna of the State and Influence of Climate and Environment on their Development.* New York: Macmillan, 1932.

Small, John Kunkel. *Eden to Sahara.* Lancaster, Pennsylvania: Science Press, 1929.

Small, John Kunkel. "Explorations in the Everglades and the Florida Keys." *Journal of the New York Botanical Garden* (No. 111, March, 1909), (No. 172, April, 1914).

Small, John Kunkel. *The Ferns of Florida.* Lancaster, Pennsylvania: Science Press, 1931.

Small, John Kunkel. *The Shrubs of Florida.* New York, 1913.

Small, John K. "The Coconut Palm." *Journal of the New York Botanical Garden* (No. 30, 1929).

Stephen, L. L. "Historic and Economic Aspects of Drainage in the Florida Everglades." *The Southern Economic Journal* X (No. 3, 1944).

Stewart, John T. *Report on Everglades Drainage Project.* Washington, D.C., U.S. Dept. of Agriculture, Soil Conservation Service, 1907.

Stockbridge, Frank Parker, and Perry, John Holliday. *Florida in the Making.* New York: The deBower Publishing Co., 1926.

Tebeau, Charlton W. *Florida's Last Frontier.* Rev. ed. Coral Gables, Fla.: Univ. of Miami Press, 1966.

Tebeau, Charlton W. *A History of Florida.* Coral Gables, Fla.: Univ. of Miami Press, 1971.

Tebeau, Charlton W. *Man in the Everglades: 2000 Years of Human History in the Everglades National Park.* Rev. ed. Coral Gables, Fla.: University of Miami Press, 1968.

Trapp, Mrs. Harlan. *My Pioneer Reminiscences.* Washington, D.C.: Graphic Arts Press, 1940 .

Troetschel, Henry, Jr. "John Clayton Gifford, An Appreciation." *Tequesta* 1 (No. 10, 1950): 35-47.

"What Would You Do? " A reporter's version of one of Dr. Gifford's stories. University of Miami. *Bulletin No. 2, Gifford Society of Tropical Botany* (1951-1952):24.

Ziegler, E. A. "We Present: John C. Gifford." *Journal of Forestry* 45 (1947):455.

Bibliography of John C. Gifford's Writings

The following sources were used in compiling the Gifford bibliography: Dr. Gifford's scrapbooks; files of *The American Eagle*

(1925-1948) from the Historical Association of Southern Florida; *The Agricultural Index;* Botany index and catalog files in the National Agricultural Library, Washington, D.C.; *Index to Publications,* U.S. Department of Agriculture; *Bibliography of Agriculture,* 1942-1949; *Reader's Guide to Periodical Literature,* 1890-1899; *Proceedings of the Florida State Horticultural Society* (1892-1917, 1920-1949); and *Tequesta,* Journal of the Historical Association of Southern Florida, 1944-1949.

"Adventitious Buds." *American Eagle* 32 (No. 21, Sept. 23, 1937).

"Aloes."*American Eagle* 35 (No. 46, Mar. 13, 1941).

"Altakapas Country, La." *Science* 20 (Dec. 30, 1892): 372

"Answering Inquiries on Trees." *American Eagle* 38 (No. 21, Sept. 16, 1943).

"Apples of Sodom." *American Eagle* 30 (No. 11, July 18, 1935).

"Australian Pine, The." *Everglade Magazine,* Jan. 1912. Reprinted in *The Everglades and Southern Florida.*

"Australian Pines in South Florida." *American Eagle* 32 (No. 30, Nov. 25, 1937).

"Back to Land Program, a Practical Solution." *American Eagle* 36 (No. 11, July 10, 1941). Reprinted from the *Christian Science Monitor.*

"Back of Fort Myers." *American Eagle* 31 (No. 14, August 16, 1936).

"Balsam Apple, The." *American Eagle* 35 (No. 33, Dec. 12, 1940).

"Bamboo and its Uses." *American Eagle* 31 (No. 37, Jan. 14, 1937).

"Banana and the Pawpaw, The." *Gardening Monthly,* August 1910. Reprinted in *The Everglades and Southern Florida.*

"Barrels, Baskets and Buckets." *American Eagle* 31 (No. 46, Mar. 18, 1937).

"Barrett's Tropical Crops." *American Eagle* 35 (No. 31, Nov. 28, 1940). (Book Review)

"Bay Rum, a Tropical Product." *American Eagle* 32 (No. 12, July 22, 1937).

"Best of All Tropical Fruits." *Gardening Monthly* 13 (Feb. 1911): 189.

Billy Bowlegs and the Seminole War. Coconut Grove, Fla.: Triangle Co., 1925. (This book consists of a reprint of a *Harper's Weekly* article, with notes and comments by JCG.)

"Bitter Barks and Their Uses." *American Eagle* 31 (No. 11, July 16, 1936).

"Bitter Barks." *American Eagle* 35 (No. 45, Mar. 6, 1941).

"Blue Mahoe, The." *American Eagle* 36 (No. 12, Jul. 17, 1941).

"Broken Pots of an Ancient Race." *American Eagle* 31 (No. 36, Jan. 7, 1938).

"Browsing vs. Pasturing." *American Eagle* 35 (No. 19, Sept. 5, 1940).

"Bucida and Buttonwood." *American Eagle* 32 (No. 36, Jan. 6, 1938).

"Bucida, a Seaside Shade Tree." *American Eagle* 32 (No. 23, Oct. 7, 1937).

"Bungalow Construction in South Florida." *The Everglade Magazine* circa 1910). Reprinted in *The Everglades and Southern Florida.*

"Bungalows for Florida." *Country Life* 19 (Feb. 15, 1911):326.

"Burial Pots of an Ancient Race." *American Eagle* 31 (Jan. 7, 1937).

"Cajeput Makes Good, The." *American Forests* 46 (Mar. 1940): 127-28. Reprinted in *American Eagle* 34 (No. 45, Mar. 7, 1940).

"Cajeput Tree in Florida, The." *American Eagle* 32 (No. 29, Nov. 18, 1937).

"Camphor." *American Eagle* 35 (No. 49, Apr. 3, 1941).

"Camphor and the Cajeput, The." *Everglade Magazine.* Circa 1912.

"Canal Trip Across Florida." *American Eagle* 37 (No. 38, Jan. 14, 1943). (Originally published in *Everglade Magazine.*)

"Candlenut vs Tung Oil." *American Eagle* 31 (No. 40, Feb. 4, 1937).

"Canella." *American Eagle* 35 (No. 43, Feb. 20, 1941).

"Cascarilla." *American Eagle* 35 (No. 44, Feb. 27, 1941).

"Castor Bean, The." *American Eagle* 33 (No. 14, Aug. 4, 1938).

"Catha." *American Eagle* 35 (No. 47, Mar. 20, 1941).

"Causes and Effects of Great Forest Fires." *Review of Reviews* 10 (Nov. 1894): 514-15.

"Change Not Always Progress." *American Eagle* 33 (No. 21, Sept. 22, 1938).

"Charles Sumner Dolley." *Fairchild Tropical Garden Bulletin* 2 (No. 6, Mar. 1947).

"Chaw-stick (*Gouaria lupuloides*), The." *American Eagle* 35 (No. 27, Oct. 31, 1940).

"Chikika, Last Chief of the Caloosas." *American Eagle* 31 (No. 38, *Jan. 21, 1937).*

"Cigar Box Cedar." *American Eagle* 32 (No. 5, June 3, 1937).
"Coco Palm, The." *Gardening Monthly,* Nov. 1910. Reprinted in *The Everglades and Southern Florida* 1911, 2nd edition, 1912.
"Coffee and Vanilla in South Florida." *The Everglade Magazine,* Nov. 1911. Reprinted in *The Everglades and Southern Florida.*
"Common Backyard Medicines." *American Eagle* 32 (No. 24, Oct. 14, 1937).
"Competition and Cooperation in Nature." *Organic Gardening* 11 (No. 3. Sept. 1947): 40-42.
"Comptie, a Valuable Wild Plant." *American Eagle* 31 (No. 9, July 2, 1936).
"Conservation of Soil Fertility." *American Eagle* 30 (No. 35, Jan. 2, 1936).
"Control and Fixation of Shifting Sands, The." *Engineering Magazine,* Jan. 1898.
"Controlled Timber Cutting." *American Eagle* 32 (No. 52, Apr. 28, 1938).
"Curiosities of Scientific Names." *Science* 22 (Aug. 1, 1939):116-117.
"Dangers of Drainage." *American Eagle* 32 (No. 11, July 15, 1937).
"Distribution of the White Cedar in New Jersey." *Gardening and Forest* 11 (No. 63, 1896).
"Distribution of Products Should Be Free." *American Eagle* 33 (No. 37, Jan. 12, 1939).
"Dwarf mistletoe, *Razoumofskya pusilla.*" *Plant World* 4 (1901): 149-150.
"Elderberry, The." *American Eagle* 35 (No. 39, Jan. 23, 1941).
"Everglade Sanitation." *Everglade Magazine,* Oct. 1911. Reprinted in *The Everglades and Southern Florida.*
The Everglades and Southern Florida. Miami: Everglade Land Sales Co., 1911. Second edition, 1912.
"Everglades of Florida, The." *Southland Magazine,* 1910. Reprinted in *The Everglades and Southern Florida.*
"Everglades of Florida and the Landes of France, The." *Conservation Magazine,* 1909. Reprinted in *The Everglades and Southern Florida.*
"Everything is Relative." *American Eagle* 42 (No. 32, Nov. 27, 1947).
"Farm Forestry." *Proc. of the Florida State Horticultural Society,* 1936: 113-116.

Fenugrek, a Medical and Forage Plant." *American Eagle* 35 (No. 35, Dec. 26, 1940).

"Fenugrek." *Organic Gardening* 11 (No. 6, Dec. 1947):50-51.

"Filibert Roth, a Letter of Appreciation from an Old Friend and Colleague." *The Forester* circa 1924. Clipping from JCG scrapbook.

"Five Plants Essential to Indians and Early Settlers of Florida." *Tequesta* 1 (No. 4, 1944).

"Fireproof and Windproof Trees." *American Eagle* 32 (No. 43, Feb. 24, 1935).

"Florida Atoll, A." *American Eagle* 31 (No. 24, Oct. 15, 1936).

"Florida Has Biggest Clam Bar." *American Eagle* 31 (No. 20, Sept. 17, 1936).

"Florida Has Diversity of Soils." *American Eagle* 33 (No. 25, Oct. 20, 1938).

"Florida Keys." *National Geographic* 17 (Jan. 1906):5-16.

"Florida Leads in Grapefruit." *American Eagle* 31 (No. 30, Nov. 26, 1936).

"Florida Mahogany." *American Eagle* 35 (No. 18, Aug. 29, 1940).

"Florida of Long Ago." *American Eagle* 31 (No. 31, Dec. 3, 1936).

"Florida of Long Ago." *American Eagle* 33 (No. 23, Oct. 6, 1938).

"Florida Owes Much To Dr. Perrine." *American Eagle* 29 (May 17, 1934).

"Florida Pencil Cedar." *American Eagle* 32 (No. 1, May 6, 1937).

Florida Trees. Coconut Grove, Florida: Forestry Dept., Florida Federation of Women's Clubs, 1909.

Floridan Keys with Special Reference to Soil Productivity. Tallahassee, Florida: Dept. of Agriculture, New Series No. 77, 1935. Reprint 1946.

"Fly Dopes." *American Eagle* 31 (No. 27, Nov. 5, 1936).

"Following an Ancient Trail." *American Eagle* 31 (No. 4, May 28, 1936).

"Forest Floss Has Market Value." *American Eagle* 31 (No. 49, April 8, 1937).

"Forest Gardening." *Organic Gardening* 9 (No. 3, Aug. 1946): 49-50.

"Forest Gardening." *American Eagle* 41 (No. 39, Jan. 16, 1947).

"The Forestal Conditions and Silvicultural Prospects of the Coastal Plain of New Jersey." In *Annual Report of the State Geologist of New Jersey for 1899,* pp. 235-318. Trenton, N.J.: J. J. MacCrellish & Quigley, 1900.

"Forestry in High School Instruction." *School Review* 5 (Nov. 1901): 560-565.

"Forestry on Sandy Soils." *Annual Report,* Commissioner of Fisheries, New York 1896: 396-417.

"Fruit Quality in South Florida." *Everglade Magazine,* March 1911. Reprinted in *The Everglades and Southern Florida.*

"Garlic." *American Eagle* 35 (No. 41, Feb. 6, 1941).

"Growing Pumpkins on Trees." *American Eagle* 34 (No. 27, Nov. 2, 1939).

"Guava and its Uses, The." *American Eagle* 31 (No. 32, Dec. 10, 1936).

"Guava and the Rose Apple, The." *Garden Monthly,* June 1911. Reprinted in *The Everglades and Southern Florida.*

"Gumbo Limbo, The." *Everglade Magazine,* Jan. 1912. Reprinted in *The Everglades and Southern Florida.*

"Gumbo Limbo Tree, The." *American Eagle* 31 (No. 39, Jan. 28, 1937).

"Historic South Florida Well, A." *American Eagle* 30 (June 13, 1935). Reprinted from the *Miami Herald.*

"Home Orchard Plan with a List of the Principal Fruits Alphabetically Arranged, A." *Everglade Magazine,* Dec. 1911. Reprinted in *The Everglades and Southern Florida.*

"How to Get a Lot of Work Out of a Small Windmill." *Everglade Magazine,* 1911. Reprinted in *The Everglades and Southern Florida.*

"Humble Koonti, The." *Gardening Monthly* 14 (Jan. 1912):276. Reprinted in *The Everglades and Southern Florida.*

"Indian Relics in South Jersey." *Science* 22 (Sept. 1, 1893): 113-114.

"Introduction of Foreign Species." *Science* 20 (Nov. 25, 1892):304.

"Karifs and Insular Caribs." *Science* 23 (Jan. 26, 1894):45-46.

"Knights of Malta, The." *American Eagle* 32 (No. 32, Dec. 9, 1937).

"Lablabs, The." *American Eagle* 35 (No. 51, Apr. 17, 1941).

"Leaching." *Organic Gardening* 10 (No. 3, Feb. 1947): 49-50.

"Leaf Fall." *Organic Gardening* 13 (No. 4, Oct. 1948): 46-48.

"Leaves for Food." *American Eagle* 35 (No. 50, Apr. 12, 1941).

"Letters from Cuba." Originally published in the *Miami Herald.* Part I,

American Eagle 32 (No. 17, Aug. 26, 1937); Part II, *American Eagle* 32 (No. 18, Sept. 2, 1937); Part III, *American Eagle* 32 (No. 19, Sept. 9, 1937); Part IV, *American Eagle* 32 (No. 20, Sept. 16, 1937).

"Licorice." *American Eagle* 35 (No. 36, Jan. 2, 1941).

"Lignum-Vitae, the Tree of Life." *Scientific Monthly* 49 (July 1939): 30-32.

"Lignum Vitae, Tree of Life." *American Eagle* 34 (No. 10, July 6, 1939).

"Lime, The." *American Eagle* 31 (No. 7, June 18, 1936).

"Lime and Sapodilla, The." *Gardening Monthly,* Sept. 1910. Reprinted in *The Everglades and Southern Florida.*

"Little Known Terms Defined (Glossary)." Part I, *American Eagle* 38 (No. 12, July 15, 1943). Part II, *American Eagle* 38 (No. 13, July 22, 1943).

"Live Oak is Strongest Wood." *American Eagle* 31 (No. 22, Oct. 1, 1936).

Living by the Land. (Conservation with Special Reference to Proper Use of the Land. A Discussion of the Conservational and Farming Methods Best Suited to the Particular Needs of Florida, the Southern United States, and the Caribbean Region.) Miami: Glade House, 1945.

"Locust Tree, The." *The Forester* 2 (1896):37-39.

"Luquillo Forest Reserve, Porto Rico, The." Washington, D.C.: Dept. of Agriculture, Bureau of Forestry, *Bulletin No. 54,* 1905.

"Luquillo Forest Reserve, Porto Rico, The." *Forestry and Irrigation* 9 (1903):537-541.

"Mahogany in South Florida and the West Indies." *Woodcraft Magazine,* August 1909. Reprinted in *The Everglades and Southern Florida.*

"Mango Known to Ancients." *American Eagle* 32 (No. 8, June 24, 1937).

"Mango, the Best of All the Tropical Fruits, The." *Gardening Monthly,* Feb. 1911. Reprinted in *The Everglades and Southern Florida.*

"Medicinal Properties of Fruit." *American Eagle* 35 (No. 23, Oct. 3, 1940).

"Men of Iron and Ships of Wood." *American Eagle* 39 (No. 9, June 22, 1944).

"The Milpah and Ejido." *Organic Gardening* 9 (No. 2, July 1946): 43-47.

"Mission of Dinner Key, The." *American Eagle* 28 (No. 49, April 12, 1934).

"More About Weeds." *American Eagle* 33 (No. 9, June 30, 1938).

"More Remarkable Koontie Facts." *Gardening Monthly* 15 (March 1912):124.

"Multiple Uses of the Guava." *Gardening Monthly* 13 (June 1911): 306.

"Names of Mahogany." *Forestry and Irrigation* 14 (1908): 196-198.

"Natal Grass a Soil Builder." *American Eagle* 31 (No. 50, Apr. 15, 1937).

"Native Tea." *American Eagle* 31 (No. 15, Aug. 13, 1936).

"Natural Resources of Florida." Eight-part series covering Florida'a Geographical Location, Fruits, Minerals, Animals, Vegetable Products, published prior to the National Conservation Exposition held in Knoxville, Tenn., October 1913. *Everglade Magazine,* circa 1912-1913.

"Natural Resources of Florida." *American Eagle* 38 (No. 14, July 29, 1943).

"Natural Sources of Fertility in South Florida." *Miami Chamber of Commerce Bulletin I* (No. 5, 1915).

"Nature Supplies Man's Needs." *American Eagle* 32 (No. 7, June 17, 1937).

"New Roots for Old Trees." *Everglade Magazine,* June 1911. Reprinted in *The Everglades and Southern Florida.*

"Notes collected during a visit to the forests of Holland, Germany, Switzerland and France." *Annual Report of the State Geologist of New Jersey,* 1896: 337-365.

"Olive does not Fruit in Florida." *American Eagle* 31 (No. 41, Feb. 11, 1937).

"Our Native Caribbean Pine." *American Eagle* 32 (No. 28, Nov. 11, 1937).

"Origin of the Word 'Key.' " *American Eagle* 37 (No. 19, Sept. 3, 1942).

"Patrick Matthew, the Forester and Author of *Naval Timber and Arboriculture* in Which He Propounded the Theory of Evolution in 1831." *Journal of Forestry,* 22 (No. 8, Dec. 1924).

"Pharaoh's Fig." *American Eagle* 31 (No. 34, Dec. 24, 1936).

"Pineapples Are Big Florida Crop." *American Eagle* 31 (No. 28, Nov. 12, 1936).

"Plant Produces Butter Color." *American Eagle* 31 (No. 48, Apr. 1, 1937).

"Pot Holers and Rock Pilers." *Organic Gardening* 10 (No. 1, Dec. 1946):23-26.

Practical Forestry. New York: D. Appleton & Co., 1902.

"Preliminary Report on Forest Conditions of South New Jersey, A." *Annual Report of the State Geologist of New Jersey,* 1894:245-286.

"Primitive Man Lived in Forest." Part I, *American Eagle* 31 (No. 42, Feb. 18, 1937); Part II, *American Eagle* 31 (No. 43, Feb. 25, 1937).

"Problem of Growing Pineapples for Market, The." *Gardening Monthly* 12 (Jan. 1911): 273-274. Reprinted in *The Everglades and Southern Florida.*

"Proper Professional Title for Foresters, The." *The Forester* 7 (No. 6, June 1901).

"Puckery Persimmon, The." *American Eagle* 31 (No. 13, July 30, 1936).

"Qualities of the Cajeput Tree." *American Eagle* 32 (No. 25, Oct. 21, 1937).

"Railroads and Forestry." *World's Work* 5 (April 1903): 347-355.

"Rare Cuban Tree in Florida." *Country Life* 25 (March 1914): 130.

Reclamation of the Everglades with Trees. New York-Boston: Books Inc., 1935. Bound with *Rehabilitation of the Floridan Keys.*

"Recollections of a Faculty Member of the First College of Forestry." *Proceedings of the Society of American Foresters* 2 (1949): 333-335.

Rehabilitation of the Floridan Keys. New York-Boston: Books Inc., 1934. Bound with *Reclamation of the Everglades with Trees.*

"Roads, Yesterday and Today." *American Eagle* 33 (No. 2, May 12, 1938).

"Romance of the Mangrove, The." *American Eagle* 31 (No. 18, Sept. 3, 1936).

"Rosary Pea, The." *American Eagle* 32 (No. 13, July 29, 1937).

"Royal Poinciana, The." *American Eagle* 38 (No. 18, Aug. 26, 1943).

"Rubber in South Florida." *Everglade Magazine,* Feb. 1911. Reprinted in *The Everglades and Southern Florida.*

"Rubber, Its History and Uses." *American Eagle* 39 (No. 6, June 1, 1944).

"Rule of Gold vs. Golden Rule." *American Eagle* 33 (No. 20, Sept. 15, 1938).

"Salt Tide Marshes of South Jersey." *Science* 22 (Sept. 29, 1893):174-175.

"Sapodilla, The." *American Eagle* 31 (No. 17, Aug. 27, 1936).

"Sarsaparilla." *American Eagle* 35 (No. 48, Mar. 27, 1941).

"Sassafras Valued by Indians." *American Eagle* 31 (No. 26, Oct. 29, 1936).

"Sea to the South of Us, The." *American Eagle* 33 (No. 12, July 21, 1938).

"Seminoles of Mixed Blood." *American Eagle* 31 (No. 19, Sept. 10, 1936).

"Septic Tanks Required for Sanitary Purposes." *Everglade Magazine,* circa 1912. (From JCG's Scrapbook.)

"Shade for Tropical Fruits." *Everglade Magazine,* Feb. 1912. Reprinted in *The Everglades and Southern Florida.*

"Shade in Southern Pastures." *Proceedings of the Fifth Southern Shade Tree Conference.* 1942.

"Silvicultural Prospect of the Island of Cuba, The." *The Forester* 6 (No. 8, August 1900).

"Sisal and the Perrine Grant." *American Eagle* 31 (No. 35, Dec. 31, 1936).

"Some Common Florida Plants." *Everglade Magazine,* circa 1910. Reprinted in *The Everglades and Southern Florida.*

"Some Reflections on the Florida of Long Ago." *Tequesta* 1 (1946).

"Some Suggestions for Future Agricultural Development of the Region around Miami." *Miami Chamber of Commerce Bulletin* 1 (No. 4, May 1915).

"Some Valuable Bean Trees." *American Eagle* 31 (No. 47, Mar. 25, 1937).

"Sources of Soil Fertility." *American Eagle* 33 (No. 5, June 2, 1938).

"Sours and Dillies." *Gardening Monthly* 12 (Sept. 1910): 62-63.

"South Florida and Panama." *Everglade Magazine,* circa 1912.

"South Florida's Resources." *American Eagle* 30 (No. 14, Aug. 8, 1935).

"South Florida, the White Man's Tropics." *American Eagle* 30 No. 26, (Oct. 31, 1935).

"Southern Florida, Notes on the Forest Conditions of the Southernmost part of this Remarkable Peninsula." *Forestry and Irrigation Magazine* 10 (1904): 406-413. Reprinted in *The Everglades and Southern Florida.*

"Sponges, a Great Florida Asset." *American Eagle* 32 (No. 14, August 5, 1937).

"Squill Valued as Rat Poison." *American Eagle* 35 (No. 38, Jan. 16, 1941).

"Strangler Figs, The." *American Eagle* 32 (No. 33, Dec. 16, 1937).

"Strangler Figs, The." *American Eagle* 39 (No. 10, June 29, 1944).

Ten Trustworthy Tropical Trees. Emmaus, Pennsylvania: The Gardeners' Book Club, Organic Gardening Publishers, Series 2, No. 7, 1946.

"Tree Roots Build Soil." *American Eagle* 35 (No. 16, Aug. 15, 1940). Reprinted from the *Miami Herald.*

"Trees of South Florida," Part I, "Five Naturalized Exotic Forest Trees: Candlenut, Sapodilla, Lebbek, Casuarina, Cajeput." *Scientific Monthly* 59 (July 1944): 21-28; Part II, Five Native Cabinet Woods: Lysiloma, Mahogany, Fishpoison Tree, False-Mastic, Lignumvitae)." *Scientific Monthly* 59 (August 1944): 101-107.

"Trees as an Aid to Drainage." *La Hacienda Magazine,* 1911. Reprinted in *The Everglades and Southern Florida.*

"Tropical Farming." *American Eagle* 41 (No. 41, Jan. 30, 1947).

"Tropical Forestry." *Modern Mexico Magazine,* October 1907.

"Tropical Nuts." *American Eagle* 31 (No. 52, April 29, 1937).

Tropical Plantation Colony on Elliott's Key, A. (Pamphlet, self-published. No date, but before 1920.) Copy in JCG Scrapbook.

Tropical Subsistence Homestead, The: Diversified Tree Crops in Forest Formation for the Antillian Area. New York-Boston: Books, Inc., 1934.

"Tropical Research Needed." *American Eagle* 34 (No. 9, June 29, 1939).

"Tropics Ideal Place to Live." *American Eagle* 31 (No. 44, Mar. 4, 1937).

"Turpentining Trees Wasteful." *American Eagle* 31 (No. 51, Apr. 22, 1937).

"Two Wonder Fruits of the Tropics." *Gardening Monthly* 12 (Aug. 1910):13-14.

"Upper Keys: the Playground of Presidents, The." (Pamphlet) Coconut Grove, Fla.: Upper Keys Improvement Assoc., 1929.

"Valuable Trees for the Everglades." *The Everglade Magazine,* circa 1910. Reprinted in *The Everglades and Southern Florida.*

"Vegetable Soaps and Dish Rags." *American Eagle* 32 (No. 15, Aug. 12, 1937).

"Vines for Everglade Planting." *Everglade Magazine,* circa 1910. Reprinted in *The Everglades and Southern Florida.*

"We Should Know Florida." *American Eagle* 30 (No. 48, Apr. 2, 1936).

"Weeds, Harmful and Otherwise." *American Eagle* 33 (No. 6, June 9, 1938).

"Weeds May Prove Valuable." *American Eagle* 35 (No. 40, Nov. 21, 1940).

"What is Muck? " *Everglade Magazine,* July 1911. Reprinted in *The Everglades and Southern Florida.*

"What the Cocopalm Means to the Tropics." *Gardening Monthly* 12 (Nov. 1910): 176-177.

"What Will Grow in the Everglades." *Everglade Magazine,* circa 1910. Reprinted in *The Everglades and Southern Florida.*

"Who Was Mr. Chi? " *American Eagle* 30 (No. 51, Apr. 23, 1936).

"Wild Tamarind, The." *American Eagle* 31 (No. 16, Aug. 20, 1936).

"Wood Useful in Button Making." *American Eagle* 32 (No. 10, July 8, 1937).

"World-Famed Forest of Vallombrosa." *The Forester,* 5 (No. 6, June 1899).

"World's Most Famous Trees." *American Eagle* 32 (No. 26, Oct. 28, 1937).

"Yaupon, The." *American Eagle* 35 (No. 40, Jan. 30, 1941).

Index